Introduction to

COMMUNICATION COMMAND
AND CONTROL SYSTEMS

Other Pergamon titles of interest

ANAND	Introduction to Control Systems
DADDA & PELLEGRINI	Automation and Instrumentation
NOTON	Introduction to Variational Methods in Control Engineering
NOTON	Modern Control Engineering
GEORGE	An Introduction to Computer Programming
GEORGE	An Introduction to Digital Computing
GEORGE	A Survey of Digital Computing
GEORGE	Computer Arithmetic
HYNDMAN	Analog and Hybrid Computing
MARTIN	Mathematics for Engineering Technology and Computing Science
WASS & GARNER	Introduction to Electronic Analogue Computers
WELLS	Elements of Combinatorial Computing
BERNSTEIN	Computers in Public Administration: An International Perspective
BULGAKOV	Electronic Automatic Control Devices
BÜRGER	Four-Language Technical Dictionary of Data Processing, Computers and Office Machines
DUMMER *et al.*	Banking Automation

Introduction to
COMMUNICATION COMMAND AND CONTROL SYSTEMS

by

DAVID J. MORRIS

*Senior Lecturer in the Electrical Engineering Department
at the Ben-Gurion University of the Negev (Israel)*

PERGAMON PRESS

OXFORD · NEW YORK · TORONTO · SYDNEY
PARIS · FRANKFURT

	4.11	Synchronous Transmission	78
	4.12	Line Disturbances	81
	4.13	Handshaking Between Modems	88
	4.14	Concluding Comments	91
	4.15	References	93

5. *Multiplexors and Concentrators* | 95

	5.1	Introduction	95
	5.2	Frequency-division Multiplexors	96
	5.3	Time-division Multiplexors	98
	5.4	Time-division Multiplexors Versus Frequency-division Multiplexors	100
	5.5	The Concept of the Concentrator	101
	5.6	Asynchronous Time-division Multiplexor	106
	5.7	Statistical Multiplexor	109
	5.8	Programmable Concentrators	110
	5.9	Concluding Comments	112
	5.10	References	112

6. *Switching Centres* | 115

	6.1	Introduction	115
	6.2	Message Switching Versus Circuit Switching	118
	6.3	Message Switching	121
	6.4	Facilities Provided by Message Switching	124
	6.5	Message Switching–Centre Structure	125
	6.6	The Functional Characteristics	128
	6.7	References	131

7. *Communication Network Hierarchy, Architecture and Control* | 133

	7.1	Introduction	133
	7.2	Network Architecture	134
	7.3	Multidrop Network Configuration	138
	7.4	Multidrop Network Design Considerations	139
	7.5	Line-control Procedures	144
	7.6	Polling and Selecting	145
	7.7	Design Concepts for Optimizing Communication Networks	149
	7.8	Alternative Routings	154
	7.9	References	157

8. *Loop Transmission* | 159

	8.1	Introduction	159
	8.2	Basic Concepts of the Loop Transmission	160
	8.3	Loop System Operation	163
	8.4	Loop Networks	167
	8.5	Alternative Routes	169
	8.6	References	172

9. *Computers in Command and Control Systems* | 175

	9.1	Introduction	175
	9.2	Time Sharing	177
	9.3	Job Scheduling	180
	9.4	Memory Hierarchy	183
	9.5	Multiprocessing System Configuration	186
	9.6	Modular Memory Organization	189
	9.7	Hierarchy of Modular Computer Configuration	193
	9.8	References	198

10. *Distributed Computer Resources* 199

10.1 Introduction 199
10.2 The Impact of Shared Resources 201
10.3 Communication Interface Hierarchy 203
10.4 Computer to Computer Communication 207
10.5 System Networks 209
10.6 References 214

11. *Terminals and Displays* 217

11.1 Introduction 217
11.2 Remote Terminals 219
11.3 Distributed Intelligence 221
11.4 Large-screen Displays 224
11.5 Large-screen Display Techniques 228
11.6 References 231

12. *Error Control* 233

12.1 Introduction 233
12.2 Error-control Parameters 237
12.3 Block Length Error Detection 239
12.4 Modulo 2 (mod-2) Arithmetic 241
12.5 Parity Error-control Codes 244
12.6 Hamming Error-correcting Codes 247
12.7 Convolutional Error-control Codes 250
12.8 Generating Convolutional Parity Digits 254
12.9 Cyclic Codes Basic Properties 257
12.10 Cyclic Error-detection Codes 260
12.11 References 261

13. *Secrecy, Security and Privacy* 267

13.1 Introduction 267
13.2 System Threat Points 268
13.3 Entrance Keys and Passwords 273
13.4 File Security and Secrecy 275
13.5 Hardware and Software Precautions 277
13.6 Cyphering Techniques 278
13.7 Concluding Comments 282
13.8 References 283

14. *Reliability and Maintainability* 285

14.1 Introduction 285
14.2 Redundancy Techniques 286
14.3 Protective Redundancy 293
14.4 Practical Considerations 295
14.5 Communication Maintainability Enhancement 297
14.6 References 300

Name Index 301

Abbreviations Index 303

Subject Index 305

Preface

This book is in essence an introduction to the highly complex and sophisticated subject of Communication Command and Control, and is offered as an initial guide to prospective system designers and the management personnel of organizations planning the adoption of these systems.

The work is concerned with both theory and practice. It approaches the subject by outlining the nature and scope of communication in general and discusses the aspects of computer techniques directly related to it. Its main purpose is to provide a sound knowledge of the basic features involved in the design of Command and Control systems, with particular attention to the communication aspects of the subject. The suggested design theory lays stress rather on the logical design of the complete system than on the planning of particular subsystems. In other words, the book is concerned mainly with the general aspects of design construction and does not purport to furnish the designer with fully prescribed mathematical tools.

A book of this nature could be expanded to cover additional topics closely or loosely connected with the general subject, but for practical reasons the presentation has been restricted to those features which have proved to be essential for the proper implementation of Communication Command and Control systems.

It must be remarked that the design ideas presented in the book are not all claimed to be original. The author's indebtedness to the work of fellow researchers can be guaged from the list of references given at the end of each chapter.

The author wishes to record his thanks to Prof. Z. Reisel of the Weizmann Institute for his kindness in reading through the manuscript and offering his expert comments on the treatment of the subject. Thanks are also due to the Department of Electrical Engineering of the University of the Negev for their grant towards publication.

D. J. M.

CHAPTER 1

General Introduction

1.1 SYSTEM DEFINITION

The growing scope of large on-line communication systems is generating new requirements and imposing additional burdens on the computer and intercomputer technologies. Fresh systems are being developed from a variety of sources and incorporated in a single comprehensive programme. These systems are characterized by the blending of at least four separate engineering mediums: (a) sensors; (b) tele-communications; (c) displays and (d) computers, which in combination are mutually fortified.

The initial task of these systems is to establish and provide an accurate real-time information status for the problems confronting an organization which will give it the means for decision making. The decision making may be manual or automatic and may be conducted at top management level or low down in the chain of command. In airline seat reservations, for instance, the decision making is done automatically by means of a machine, while in medical diagnostics the decision making is done manually and here the machine merely assists in presenting the information status.

These systems are a complex of procedures, doctrines and devices which supply operational logistics and administrative information for the decision making and the instruments for implementing these decisions. They comprise the projects of data collection from various topological centres, the correlating of the data with previous data-processing problems, updating the files, displaying the information status, dispatching and executing the decisions. The systems are directed at large organizations to assist them by providing them with means for planning, managing and controlling their active operations, using on-line information and implementing it in real-time.

The main feature of these systems is that the whole performance is conducted in a closed loop. Here the decisions are made at the upper levels, from where there is a feed forward of commands and to which there is a constant feedback from the lower levels to indicate the execution of the commands. This provides the upper level with a real-time status picture of what is happening, thus enabling decisions to be made "on the spot" within the course of an operation. The closed-loop principle thus ensures a more efficient control of the organization resources by the management. A schematic flow-chart of such systems, which gives the essential tasks used, is shown in Fig. 1.1.

Various terms have been adopted to describe these complex systems, but the broadest and most appropriate term for these schemes is COMMUNICATION COMMAND and CONTROL, often denoted by the term C^3. A Communication Command and Control incorporates many system skills, such as telemetry, remote processing, teleprocessing and time sharing, but each one of them is only part of the techniques used in the

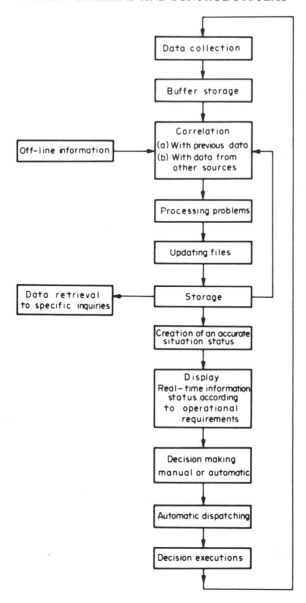

Fig. 1.1. Schematic flow chart of C³ systems.

comprehensive system; one could add a long list of further techniques which could be used. Communication Command and Control in essence is the centralization and coordination of sets of various resources which are physically remote from the centre, using all the required techniques available.

Though Communication Command and Control is a military term the systems under this title are not restricted to military applications. Far from it, they are intended for a wide range of disparate applications. A small list of some of the C³ applications is given here:

(a) Airline seat reservations.
(b) Hotel room reservations.
(c) Car-hire firms.
(d) Medical centres for automatic patient care.
(e) Air-traffic control.
(f) Electricity distribution supplies.
(g) Traffic congestion.
(h) Remote bank accounting.
(i) Stock marketing.
(j) Transportation system planning.

Some of the applications of C^3 systems have succeeded and are being extensively adopted. Where other applications have been tried and later rejected, the cause of failure may have been due either to inadequate mastery of the subject or to the slipshod design procedures employed.

A major common factor in all the diverse applications is the requirement of a centralized coordination of the resources physically located over a scattered geographical area. This implies that C^3 is directed at large organizations, to enable them to make full use of their resources and achieve extra efficiency, improved accuracy, higher reliability and a faster system reaction.

It must be emphasized from the outset that the centre of communication, command and control systems is always the MAN. Computers are used in the system but only as a tool. Though the computer can assist in assessing and appraising the information, no computer can replace human intuition.

1.2 SYSTEM DESCRIPTION

Communication Command and Control systems have been characterized as the centralized coordination of resources physically located over a scattered geographical area. They comprise data collection, processing, updating, displaying and decision making. Taking all these factors as a single entity, one can visualize a system as shown in Fig. 1.2. The schematic model given here is one which can fit many systems, such as Remote Data Processing and Time-sharing systems, where a number of remote terminals

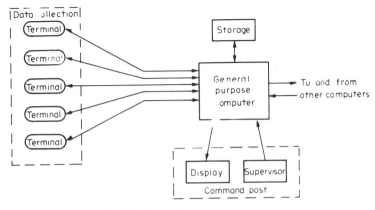

Fig. 1.2. Basic C^3 system configuration.

can communicate simultaneously with the central computer.

C^3 systems are much more complicated than the basic model and may comprise a number of computers arranged in a hierarchy of responsibilities and operating independently. The labyrinth of problems confronting the construction of the C^3 system organization is further complicated by the distance of the terminals and the computers from the centre, the addition of types to terminals, the increase in the volume of the information flow, and the diversity of the different processes required.

Communication Command and Control must be considered as a collection of systems each using different techniques and each possibly containing a computer or even a set of computers. Although each particular system may be regarded as a system in its own right, in conjunction they operate as a single unit. The main problem in this vast system is the interface between the different system techniques, for there are at least six sets of techniques which make up the communication command and control system:

   (a)   Data collection systems (sensor base system),
   (b)   Data execution systems (transaction system),
   (c)   Display systems,
   (d)   Command supervisory systems,
   (e)   Communication network systems,
   (f)   Central processing systems.

The main concern in designing Communication Command and Control systems is not the treatment of each system separately but incorporating them into one system and ensuring their proceeding in unison. The hierarchy of these systems and the connecting lines interfacing between them are shown in Fig. 1.3.

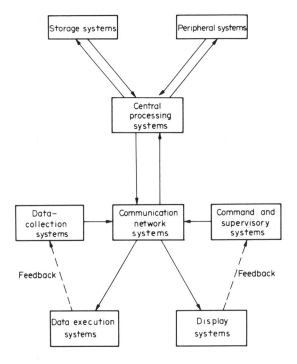

Fig. 1.3. C^3 hierarchy of systems.

It is hard to state which is the dominating factor of the C^3 system, as each technique has the same importance in the working of the whole system and the failure of one part is the failure of the whole system. From a designer's point of view the dominating factor is the communication network, whereas from a user's point of view the essential feature of the system is the command supervisory post.

The design of C^3 systems as so far discussed has regarded the comprehensive scheme as a hierarchy of system techniques, thus implying a vertical construction. It should be realized, however, that the C^3 systems may be viewed also horizontally, as a series of systems each with its own subsystems or modules. Each subsystem in turn may be

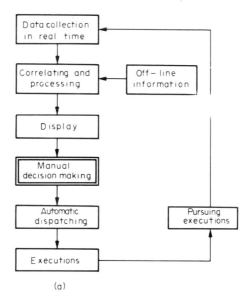

(a)

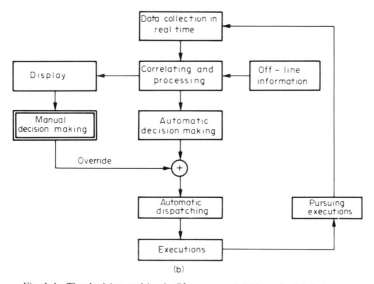

(b)

Fig. 1.4. The decision making in C^3 systems. (a) Manual. (b) Automatic.

regarded as a separate C^3 system configuration which utilizes all the resources of the techniques already discussed. Furthermore, although each subsystem may be intended for a particular operation, it may also have the use of common equipment, such as the communication network, processing or display systems. When designing a C^3 system, one should first specify the master plan for the overall comprehensive system before proceeding to deal with the subsystems in detail. The introduction of each modular subsystem into the working system may be in intervals of months or years. For this reason the design must ensure in advance that the further introduction of the subsystem will be integrated into the general system and will not involve any drastic changes.

The delay in the introduction of the other subsystems may be due to lack of immediate finances or to the need for a prior subsystem to prove itself before other subsystems are introduced.

1.3 THE COMMAND POST

The performance of Communication Command and Control systems may be divided into two types of operations (as shown in Fig. 1.4):
(a) management systems for commands, where the continuous intervention of humans is required;
(b) control systems for execution – where the decisions are made automatically, with the supervision of humans to override these decisions if necessary.

Large C^3 systems may incorporate both types of operations in the same configuration, although in different modules of the same system. Some of the subsystems of the same application structure may be of type (a) while others of type (b).

This whole performance of C^3 systems centres round the supervision post, which is termed the "command post" and is shown in Fig. 1.5. The command post is the action

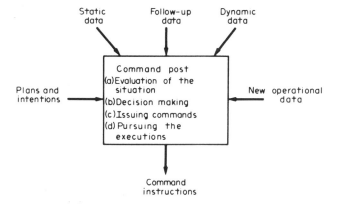

Fig. 1.5. The command post.

arm of the system, and it is here that all the relevant information is collected and displayed, on the basis of which decisions are made and commands issued. It is to be noted that the command post may be controlled by human agents or by computers.

The command post is kept supplied with moment to moment information, enabling it to adjust to the situation status. Thus equipped, the command post can perform the

following functions:
  (a) evaluation of the situation;
  (b) decision making;
  (c) issuing commands;
  (d) pursuing the execution of the commands.
For it to be able to maintain and pursue its tasks effectively, the command post must be presented with various types of information data:
  (a) static data which do not change frequently;
  (b) dynamic data which must be updated at regular and frequent intervals, or whenever there is a major change;
  (c) new operational data which must be updated in real-time and in an on-line mode;
  (d) follow-up data of the execution of the commands;
  (e) plans and intentions of the organizations.
Each Communication Command and Control system application may have a different interpretation of the items listed above. This makes it extremely difficult to include all types of applications in one context.

1.4 SYSTEM BENEFITS

Each Communication Command and Control system is intended for a specific application; hence, it is practically impossible to list together all the benefits one can achieve by designing a C^3 system. Each application could present a different target list showing the achievements which may be obtained by redesigning the current system organization. In the list presented below, an effort has been made to summarize those items which apply to most applications.
  (a) Centralizing the organization efforts in controlling and monitoring the resources.
  (b) Releasing operators from all routine and repetitive tasks.
  (c) Transferring the operators' efforts from routine to evaluation, detecting, analysing or decision making.
  (d) Expansion of service facilities.
  (e) Data collection from various remote sources at high speeds and constructing real-time situation plans.
  (f) More efficient usage of resources.
  (g) Correlation of data collected from remote sources in real-time.
  (h) Expanded service facilities.
  (i) Reducing operational errors.
  (j) Optimizing for resource management.
  (k) Less dependence on staff with long experience.
  (l) Reducing the need to rely only on the human memory and transferring all data collected to computer files.
  (m) Creation of memory files of past and present information which can be retrieved without delay.
  (n) Remote access to files in real-time.
  (o) Display of on-line real-time information for decision making.
  (p) Shortening the delay gaps between the location of problems at the remote stations and their presentation at the centre.

(q) Shortening the delay gaps between the decision making and its execution.
(r) Means for feedback control on the execution of decisions.
(s) Efficient utilization of the communication network.
(t) New and improved communication services between remote centres.
(u) Concise, accurate and timely display of alarm situations.
(v) Immediate detection of degradation of performance and service.
(w) Automatic check of faults.

1.5 SYSTEM ASSESSMENT

The development of the C^3 system is extremely expensive and its implementation may affect the working of the whole organization. So, before committing oneself to the adoption of a particular design, the cost effectiveness of the system must be considered. A number of papers have been published on the economical parameters as means of calculating the cost effectiveness. However, there is no single parameter that can be used here, as the cost effectiveness for these systems cannot be given a mathematical equation which can be generally applied. Furthermore, the actual cost of a system is not an easy figure that can be calculated, as there is not only the cost of the equipment, but also the involved cost of salaries of operators, the training of personnel, building, stoppage of production during installation, running mistakes, etc.

Let us consider the main aspects of the question of assessment:

Principal considerations

Saving of financial expenditure.
Speed of reaction obtained from the system.
Volume of data flow in the system.
Routine and parallel operations that are diverted to the machine.

What does the system save?

Money – Will there be a saving in the running cost?
People – Will there be a need for other types of people with different training?
Equipment – Will there be a need for other types of equipment at present unspecified?
Time – This is usually the main aspect which shows a pronounced saving factor.

Investment considerations

The cost of the system.
The financial payment, immediate and prospective.
When will the system be ready?
What will be the disturbances to the present system of operation?
Would it be possible to return to the original system operation?
What will be the consequences of failure?

The list mentioned above is rather vague and it is most difficult to evaluate from it the true cost effectiveness of the systems. C^3 systems are complicated and sophisticated and a parameter must be used when considering the economics of the system. A tentative scale evaluation is suggested below:

(a) *Reaction time of the system.* This is the accumulated times of lags in the system from the moment the operator presents his call for an action and the time the answer is received and implemented.

(b) *System throughput.* This is a parameter where the reaction of the system is measured in the number of messages transmitted in the system per second (or per minute).

(c) *Number of users.* This is another accepted parameter used mainly for time sharing types of systems, although it could also apply to C^3 systems. This scale is defined as the maximum number of users the system can handle simultaneously without the degregation of services.

(d) *Performance scale.* A number of system performances may be added as a scale in assessment of the virtues of C^3 systems, although they must not be considered by themselves:

1. The number of errors the system introduces.
2. The accuracy the system offers.
3. The system reliability.
4. The system dealing with overflow.
5. The system dealing with failures.

1.6 CLARIFICATION OF TERMS

There are three general terms which are constantly used in Communication Command and Control systems, but as they are also applied in other fields with possibly different interpretations, it is well to define them here.

On-line

An on-line operation is one which transmits its information about a current activity as soon as it occurs. This is made possible as the system is directly connected to the sensors and/or terminals.

The C^3 system uses on-line operations and on-line information in most but not necessarily in all of its subsystems.

Real-time

A real-time operation is one which presents an answer to a continuing problem for a particular set of input values while those values are still available.

The term real-time has been used in industry for a variety of operation interpretations. When designing a C^3 system one must define as accurately as possible what is meant by the period allowed for a real-time operation. Some systems may allow only a delay of microseconds whilst other systems may accept a delay of hours and will still call their system "real-time".

Terminal

A terminal is an input–output device situated at a point at which information can enter or leave the system. It is designed to receive data in an environment associated with the specific function performed and capable of transmitting it over a communication path.

The term 'terminal' is used throughout this book in its loose interpretation, that is, as an end node of a communication line. In this respect the term can apply to a variety of equipment, from a simple teletypewriter to a computer.

1.7 REFERENCES

1. Unold, R., "Communication Command and Control – a few worms from a can of worms", *Telecommunications* (Mar. 1968), pp. 13-16.
2. Reyes, R.V., "Real-time military communication for the 1970's", *Signal* (Oct. 1970), pp. 22-26.
3. Weiss, G., "Restraining the data monster: the next step in C^3", *Armed Forces Journal* (5 July 1971), pp. 26-29.
4. Hilsman, W.J., "The design and operation of an automated Command and Control system", *Military Review* (Feb. 1967), pp. 22-29.
5. Steinkraus, L.N., "Command and Control", *Ordinance* (Mar.-Apr. 1971), pp. 440-442.
6. Davis, C.J., "Command control and cybernetics", *Army* (Jan. 1963), pp. 51-55.
7. Gould, G.T., "Command, Control and Communication systems", *Signal* (May 1966), pp. 70-71, 79.

CHAPTER 2

System Design Concepts

2.1 THE PROBLEMS CONFRONTING THE SYSTEM DESIGNER

Before dealing with the concept of design it is worth reaffirming the general purpose of Command and Control systems, which is to provide improved control of the organization by enabling the top level management to know what is occurring in the lower levels so that they can make their decisions in time to affect the process. The design purpose of the system must therefore be to exploit all the elements involved whether they are in the upper or lower levels.

In many engineering sciences, the design is based on mathematical equations, but in Command and Control systems there are a number of hypothetical elements, such as the human factor, which cannot be formulated mathematically. While it is possible to formulate equations related to human behaviour by taking the average reaction time of a typical operation to a specific problem, this step would be of no avail to the system designer who is concerned with human reaction to problems of decision making and intuitive behaviour. The designer therefore adopts a different approach. He reduces all the hypothetical problems to a series of factors by defining the data flow in the system, showing all the possible paths or branches.

The initial stage, and perhaps the most important step, is to define the requirements of the system. In this stage the problems are defined, the applications identified and the objectives specified. Only after the problems have been properly determined, stated and dimensioned, can all the subsequent design stages be warranted. To ensure this, the designer must concentrate on these initial stages, making preliminary investigation to define the scope of the system and specifying all the contingencies which could arise for decision making.

As previously indicated, the Command and Control system is a closed-loop construction, with constant feedforward and feedback between the upper and lower levels of the organization, although for design purposes the loop may be interrupted. The system itself could be regarded as a black box (Fig. 2.1) where only the input and output are considered in the initial design stages.

Starting with the "black box" output, the designer first analyses what is required of the system, that is, what information the system has to deliver. This could be regarded as the definition of the situation status to be displayed for decision making.

The design of the black box output must determine the following points:

(a) The nature of the information that is needed.
(b) When is it needed?
(c) Who needs it?
(d) Where is it needed?

11

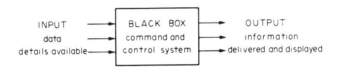

Fig. 2.1. System black box.

(e) In what form is it needed?

(f) The relative priority treatment of the information.

Once the black box output has been defined, the designer can pass on to the input side to determine what are the raw data details required for constructing the processed information. The data details available at the black box input may be far more or far less than those required to produce the status information display at the output. First, the designer must determine the data details available at the input, viz.

(a) What data is available?

(b) When can it be supplied?

(c) Who can supply it?

(d) From where is it supplied?

(e) In what form is it supplied?

(f) The relative priority of scanning the data.

After both the output and input requirements of the black box have been defined, the designer must ensure that all the data that can be collected within a given period of time is adequate and deliverable at the necessary speeds, so that the situations requiring decisions can be controlled. If the input does not satisfy the output requirements, a complete new data-input collection system must be designed.

With the system requirements defined and the input sources identified, a flow chart of the information can be outlined. This flow chart will demonstrate all the steps required for the data flow within the system, from the input terminals, via the decision post, up to the execution stations of the commands. The flow chart will show all the available paths that have to be taken to perform each decision operation.

At this stage it is desirable to carry out a thorough analysis of the system flow, using operations research or organization and methods technique. It is not satisfactory just to plot the current situation as described in the flow chart and mechanize it blindly. The current flow of information is usually of a nature and amount to suit situations which the organization was confronted with at different historical periods. (A decision operational path may pass through a department which is only loosely associated with the subject merely because the departmental head was once in charge of the subject.)

Only when the complete system is defined and new flow charts constructed can the design of the black box contents be initiated.

The next step is for the designer to systematically tabulate and analyse the operation, taking into account the following points:

(a) What were the operational problems that initiated the need for the new system? This calls for close contact with the operators and observation of the current system operation while it is in process.

(b) Would the new system solve all the revealed problems?

(c) What are the respective responsibilities and duties of the people involved in the new system design, namely, the managers, supervisors, operators and maintainers?

(d) How does the human element fit into the new system? Do the demands accord with their status and capabilities?

The points listed above are decisive and must be kept in mind throughout all the stages of design. The designer should insist on the operators of the current system associating themselves in all the stages of the design in order to help the designer in specifying the flow charts. This will enable the users (operators) to learn the system which is planned and designed for their future use. In this way the actual users could contribute to the design and suggest changes in procedures which in their experience have proved to be not feasible. The designer should encourage the prospective users to take personal interest in making the system work. Their cooperation will reduce many future contingent problems which might arise when the system is implemented.

The designer has above all to ensure that the proper emphasis has been placed on the elements of the system, in particular on the interaction of man and machine. He must take care not to add equipment for the simplicity of design rather than for the effectiveness of the system, and certainly avoid adding equipment for the sake of mere impression.

When designing a Command and Control system, the designer should be careful not to distort the basic concept by placing undue emphasis on particular features. Above all, he should refrain from regarding the mechanization or the automization of a system as the design goal, ignoring the true role of machinery and computers as tools for use in the system.

There are two important considerations which should be taken into account when evaluating and analysing a system:

(a) Not all of the systems' operations that call for modifications should be automated. There are numerous methods that may be efficiently implemented without using mechanized procedures.

(b) Not all operations that can be mechanized should be so dealt with. A thorough analysis is always needed to see whether the introduction of a computer is advantageous and really necessary for the system in question.

To sum up, ideally, the system evaluation calls for a designer who is an expert in many fields. In practice, however, the design of Communication Command and Control systems demands the services of a team of specialists each concentrating on only one aspect of design but all coordinating their activities. This team should include the following:

(a) System design engineers, under guidance to avoid creating an unbalanced system using superfluous equipment.

(b) Software system analysts, who likewise cannot be given a free hand to computerize, ad. lib. or to introduce programmes which are too sophisticated for others to handle.

(c) The eventual operators (the customers), whose elaborate requests must be carefully checked and related to their ability to handle the complete system and to the limitations of the project budget.

(d) Operation research analysts, who must not be allowed to regard the problem as predominantly an academic exercise, or to aim at an ideal but impracticable standard of efficiency.

(e) Human research engineers, who must in turn be restrained from demanding so many improvements so as to make the overall cost unduly high.

Before going into the design details, some thought must be given to the Man–Machine interaction, since one of the crucial features to be considered in C^3 system design is the involvement of the human element as an integral part of the system in operation.

2.2 MAN-MACHINE INTERACTION

In all Command and Control systems the machine (computer) appears to be the main implementing factor. This, however, is an erroneous impression. The machine is not the dominating element but rather an instrument generally employed to relieve humans of the onerous load of repetitive and routine operations and freeing them for analytical operations. In no sense is the machine intended to replace the human operator completely. For the machine to serve efficiently as an aid, the designer must know how to exploit its potentialities ever in the right direction.

The introduction of computers into Command and Control systems requires further examination, since the computers have been called intelligent machines performing sophisticated tasks while in fact they are dumb machines that only perform what they are told to do. Man-machine interaction is most important when the computers are employed for problem solving and decision making. Problem rationalization is inherently in the computer domain, while the problem formulation lies exclusively in the human sphere. For collaboration between man and computers, it is vitally important that they should be able to communicate with each other. It is necessary, therefore, to appraise the functions and nature of each, and above all their interaction. It is well to summarize where and how the sophisticated machine can assist the human operator:

(a) Releasing humans from all routine and repetitive projects and diverting their efforts to analysing and clearing ambiguities.
(b) Diverting the human analytic ability to situation synthesizing and decision making.
(c) Reducing the need for the application of human memorizing by means of a mechanical library.
(d) Reducing the load of manual operation required of humans.
(e) Increasing the individual output by enabling the human operator to deal with more items simultaneously.
(f) Reducing human errors (by transferring all the repetitive tasks to the machine).

At this point it is to be noted that although the machine assists humans in performing a variety of tasks, the introduction of machines does not necessarily reduce the number of people. A sophisticated machine system requires operators, maintainers, programmers, analysers and, above all, supervisors to perform the decision making. In fact, it has been established that the introduction of machines brings to light further tasks that can be handled by the system and consequently the number of people may even increase.

For the machine to serve efficiently as an aid, the designer of C^3 systems must know how to exploit its potentialities and limitations. It is well to draw up a table of all the relevant points which apply when discussing man-machine interactions (Table 2.1).

A good system design takes into account the complete system including all the elements involved. No system could be designed without considering the machine, in fact, no system could be designed without considering the place of MAN in the system.

TABLE 2.1

Function	Humans	Machines
Analysis ability	Sophisticated	Superficial
Reception	Picturesque configuration	Characters and bits
Means of reception	Random	Sequential
Speed of reaction	Relatively slow	Extremely fast
Description ability	General and obscure	Clear and precise
Type of reaction	Sophisticated	Strictly according to programme
Attitude to problems	Suspicious	Direct with no hesitation
Attitude to commands	Mutinous and rebellious	Obedient and passive
Proneness to error	Extremely high	Relatively low
Adjusting to new situation	Yes, after training	Only if originally considered
Deterioration with time	Prone to fatigue	No change
Reaction to environment (noise)	Reduces faculties	Completely immune

Therefore, the system designer must regard the humans as one of the elements of the system, in the same manner as he regards the machine. The humans are not only associated with machine systems but are an integral part of the overall system, with the same indispensable importance as the machine.

Without human intervention the system may not operate properly or may not operate at all. Hence the system designer should ensure that the humans are not the bottleneck of the systems. For this, means must be provided to speed up the humans' reactions. It is not only the two independent elements, man and machine, that must be considered: there is also a third element that makes up the system, and that is, the interaction itself. The actual interaction properties must always be taken into account in designing a command and control system and it is worth while to regard it as a separate feature.

It is an important design procedure to define not only how the machine assists the humans but also how the humans assist the machine. In the combined man/machine systems each side assists the other in a manner we have called interaction. In a number of sophisticated systems, the computer may literally ask the operator questions on how to continue its operations, or it may call on the operator to provide it (the computer) with further information. In this case the machine insists on the human intervention, and without it, it ceases to operate. The creation of a special language for the interaction of man and machine reduces the operational errors and ensures that both the system elements operate in a fixed high standard pattern.

No C^3 system can be fully automatic without ambiguous situations arising which consequently cause operational errors. This applies specifically to Command and Control systems where humans play a distinctive part in the operation of the system by supervising and overriding the machine decisions. The question is, what are the respective parts to be given to the humans and to the machine. There is a danger that costly human error will occur if too many tasks are given to humans. On the other hand, if too much stress is laid on the machine it will cause mechanical faults which can result in the operation being unable to deal with unforeseen situations.

The suggested design procedure (Fig. 2.2) is presented to define man-machine interaction as viewed from two angles.

(a) What key role do the humans play in establishing the ultimate operational

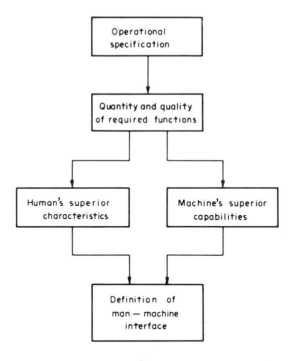

Fig. 2.2. Criterion for defining man–machine interface.

performance of the system and what tasks can the humans perform better than the machine?

(b) What part of the system is indisputedly in the machine domain and where can the machine exceed human ability?

There are a number of aspects which must be considered when designing the system which could assist in defining the borderline areas of man–machine interaction.

(a) The loads a human (or a machine) can carry.

(b) The exhaustion of humans in performing specific operations.

(c) The training time required from humans.

(d) The difference between the reaction of a diversity of individuals to the same problem.

(e) The flexibility of humans and machines in reaction to new problems.

(f) The cost of transferring responsibilities from humans to machine, and vice versa.

A most interesting example in the classification of man–machine interaction is taken from space flight. Initially, limited powers were delegated to the astronauts but today more and more responsibilities and decision making are transferred to them.

Although in the design of Command and Control systems emphasis is concentrated on the human interaction with an operating system, no system design should be considered without analysing the human engineering aspect of the system, for the humans are an integral part of the system and their physical and psychological behaviour could affect the whole system operation.

Since humans are susceptible to environmental conditions and are affected by the length of time they are associated with the operation, the suitable application of the humans to the working system is essential for the success of the system. An excellent system design which took into consideration all the requirements of human-machine interaction, but neglected the human engineering aspect of operating the system might fail if, for instance, the operator's chair proved to be uncomfortable.

Human engineering is intended to deal with all the problems of the human performance in an alien engineering environment. It is aimed at enabling humans to operate for long periods in front of the system panel in comfort and ease, without causing any undue errors.

A system performance is limited by the behaviour of its major components, namely man and machine (as shown in Fig. 2.3), while machine behaviour can be measured in a

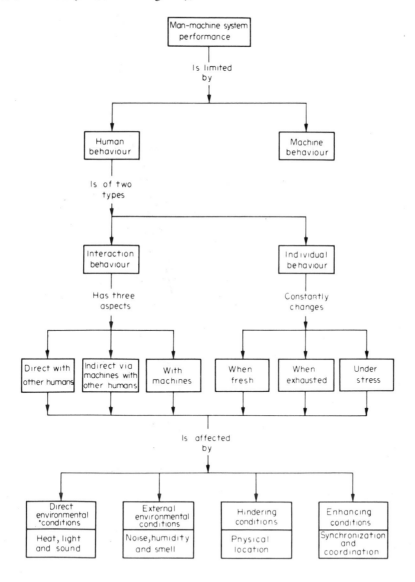

Fig. 2.3. Human behaviour in the operational system.

known manner, human behaviour is an unknown factor, as it is formed by a complex set of reactions. Human behaviour has two aspects: in isolation and in interaction with others. Humans can directly interact with other humans, indirectly with other humans via a machine or only with a machine. Individual behaviour is constantly changing and performs variably when fresh, when exhausted or under stress. Humans are affected mainly by four different conditions:

(a) Direct environmental conditions, such as heat, light and sound.
(b) External environmental conditions, such as humidity and smell.
(c) Hindering conditions which are caused by the physical location of people relative to the machine.
(d) Enhancing conditions which affect the human synchronization and coordination with the machine.

All the aspects of human behaviour relative to the system performance must be clarified and defined before the final design of the system, to ensure proper participation of the humans with the machine.

2.3 MAN VERSUS MACHINE IN THE OPERATIONAL SYSTEM

Command and control systems are man orientated, that is to say, man is an essential part of the system loop, where his main task is decision making. Sensor handling, data collection, data editing, display generating and the decision distribution can be transferred exclusively to the machine, as these actions are of a routine and repetitive type. It is the position of the humans in the decision making that is questionable. Numerous solutions have been proposed for utilizing the human element in an operational system which is based on the philosophy of the system design. Many systems have failed owing to the wrong placement of the human operators in the loop. The possible different design philosophies in loop configuration are presented here and are later analysed.

(a) Where only man can make the decision; without his intervention all work would cease and the operation be at a stand-by (Fig. 2.4).
(b) A completely automatic system where the machine carries out all the operations. In this system, the humans play no active part in the loop (Fig. 2.5).
(c) An automated system, as in (b), which operates with no human intervention. However, when there is a fault in the system, the operation is transferred to a manual backup system. In other words, when the automatic system is unable to carry out its tasks, it is transferred back to manual operation as the system was originally operated (Fig. 2.6).
(d) An automatic system where all the routine and repetitive decisions are carried out and performed by the machine. Nevertheless, all doubtful and ambiguous information must be handled by humans (Fig. 2.7).
(e) An automatic system where all routine and repetitive information is handled by the machine. Nevertheless, a modified version of all the relevant routine information is displayed in front of the humans in a summarized but precise manner. The humans supervise the running of the system without interrupting the operation; however, if they feel the smoothness is interrupted they can intervene, if necessary. The humans are presented with tools to cancel the machine decisions

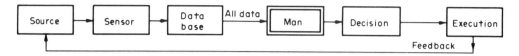

Fig. 2.4. Man-dominated system.

Fig. 2.5. Machine-dominated system.

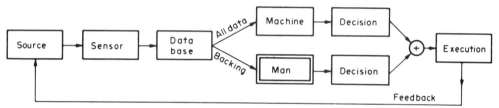

Fig. 2.6. Manual backing system.

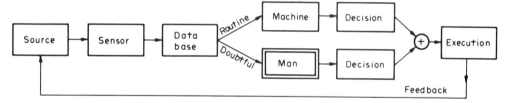

Fig. 2.7. Man assisting system.

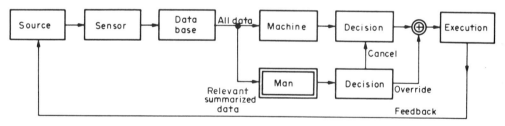

Fig. 2.8. Man supervisory system.

and to override the execution of the commands (Fig. 2.8).

For Command and Control applications, the first three types of system loop must be rejected and the designer should avoid using them. No system should be designed where only the humans or only the machines perform the decision making. Decision making is not only a matter of intuition, which is inherently in the human domain, but also based on previous experience which can be programmed into the machine.

There is no command and control system which does not contain a number of routine and/or repetitive operations which cannot be performed by the machine. On the other hand, there is no Command and Control system based only round the machine operation and relying only on its decisions. This may be right for production systems but not for Command and Control systems. Nevertheless there is always a need for some decision making based upon new data or ambiguities and problems unforeseen originally. Furthermore, the decision making is too important to rely only on the machine. Humans must supervise to verify that all the decision making is correct and really necessary.

The use of a manual standby system in the case of machine failure is a completely wrong design attitude, one which indicates that the designer is not too sure of the reliability of the machine system. If the designer bases his system design on the utilization of a machine for the routine decision making, he should rely on a second machine as a standby which is operated when the first one fails. If the designer requires greater reliability, he can add a third machine in parallel to the other two, but he should not rely on humans to take the place of machines.

In Command and Control systems the last two types of loop systems should be adopted. That is to say, the system should be fully automatic with the humans at a continuous stand-by. In complete system loops there should be three human posts, one supervising the operation, the second dealing with all doubtful information, whilst the third post concerned with data enquiries and data retrievals (Fig. 2.9).

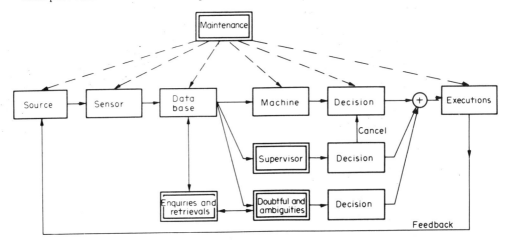

Fig. 2.9. The complete system.

The information for all these three posts is presented via the media of display systems (for this purpose, hard copy is also regarded as display). Care must be taken that only the most relevant information is displayed. Humans should not be swamped with more details than they can handle. The reaction of the humans slows down as the number of items increases. All the information which is displayed must be sensored by the machine while only a summary of the precise data should be presented.

2.4 THE HIERARCHY OF SYSTEM DESIGN

In the discussion up to this point, the system design was dealt with in general terms. The actual design will be found to be far more tedious, as it has to cover many structural

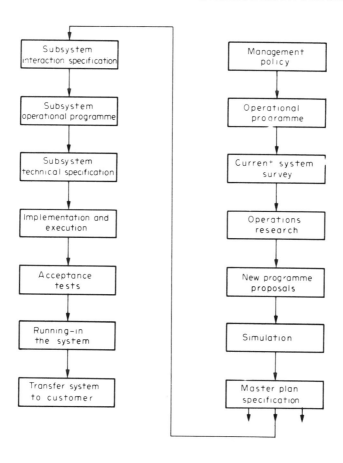

Fig. 2.10. Hierarchy of system design stages.

stages. Several of these design stages are shown in Fig. 2.10, and in the following sections some of them are expanded, demonstrating with the aid of flow charts the steps the designers must take in the definition and the identification of the problems.

Starting from the top, the first stage must be the management policy specifying the scope of the project, i.e. the part of the organization that should be investigated, the time when it is intended to operate and the amount of money that is available for the project.

The operational programme specification, being the fundamental step in the system design, must be performed within the organization itself. This is the stage for denoting the scope and requirements of the whole system. While the management may employ a consultant to advise them on what they can expect from a Command and Control system, it devolves on the management to specify the policy for themselves rather than just approve outside advice.

The operational programme is an extension of the management declaration of policy, defining in detail the requirements of the system. As shown in Fig. 2.11 the operational programme is divided into three procedures: system objectives, system scope and the fundamental postulations. Each of the procedures calls for a separate discussion.

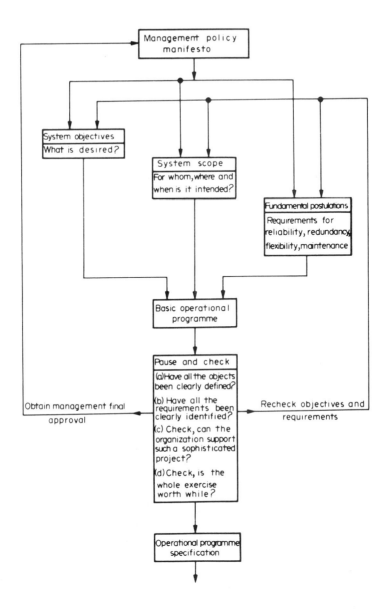

Fig. 2.11. Operational programme-design flow chart.

(a) *System objectives*

Here the management must define the objectives they expect to achieve by going in for a Command and Control system. They must specify the problems that cause them to look for new operational solutions and the performance they expect to achieve when the system is developed. The objectives are essential for the designer, and so must be clearly and accurately defined.

(b) *System scope*

With the objectives and requirements of the system defined, there is now a need to specify the scope of the project. This stage is based on the expansion programme of the organization. It is intended to clarify the approximate size of the expected system by defining whom it is intended for and the facilities available at each location. Some of the points which are to be realized within the scope procedures are given below:

1. The long- and short-range plan of the organization programmes.
2. Whom the system is intended for and who is to be associated with it.
3. Where the system is to be erected.
4. What is the initial budget (for basic design) and what is the approximate budget that will be available for the development of the complete system?
5. A general scheduling programme; when is the system required and for how many years must it remain operative before a more advanced system would be needed?

(c) *Fundamental postulations*

The objectives and scope of the system must be specified clearly and accurately so as to define the boundaries of the system. Nevertheless, there are many further items that must be postulated by the management although they cannot be accurately specified. For the postulation, the management must decide on the general lines which should be used, rather as a design guide than a design command. The designer, although allowed flexibility in his design, must obtain full management approval for any postulation changes, such as the following:

1. The modular subsystem required in the system according to their relative importance.
2. The approximate time-table for developing the subsystems (based on the organization growth programme).
3. The required reliability of the system.
4. The allowed redundancy. (How much is the management ready to pay for reliability?)
5. The maintenance required and its effect on the system operation. (Can the system or subsystem be halted for routine services?)
6. The backup required in case of total breakdown.
7. The requirement for hard copy.
8. The safety requirements for people and equipment.
9. The maximum accepted error rate.
10. The information security requirements. (Who is allowed to retrieve or update the information?)

When all the requirements of the system have been decided, a pause is necessary to check and recheck the operational programme. The programme is returned to the top management for final approval. The management is now in a position to decide whether the system is worth while and if the organization can support a project of such a scale. If the answer is negative, the project should be shelved for future reference or adoption. If the answer is favourable, the organization can instruct the design consultancy company to continue with the system design.

However tedious and unnecessary the system operational programme specification may sound it is extremely important. No organization should plan any changes without knowing what they are getting into and what these changes will mean to the organization operation, for once the development of a Command and Control system is initiated, it will tend to distort all the routine work of the organization and call for continual adjustments.

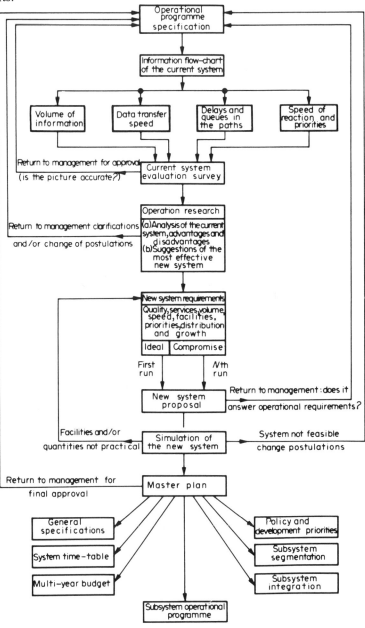

Fig. 2.12. Master plan-design flow chart.

2.5 MASTER PLAN SPECIFICATION

The operational programme specification was described as the fundamental step in the system design; however it is the master plan which is the more dominating and difficult factor. The master plan is the programme that specifies the whole system which is to be developed (Fig. 2.12). Before this can be accomplished, the current system operation is 'photographed', to enable it to be analysed and its advantages and disadvantages listed.

First, a flow-chart describing the information flow in the current system is plotted. The chart should indicate the geographical locations of each of the information sources and the intermediate stations the information passes through before it reaches its final destination. A special note should be taken of those intermediate stations where the information undergoes some kind of transformation and of those where decisions are made.

Once the flow-chart is available, a statistical investigation of the whole current system is conducted. In this investigation, the volume and speed of information flowing through the system is analysed, showing its distributions over the different available paths. The volume and speed are measured, at fixed intervals, over long periods of time. This enables the definition of the average values and the peak values to be made. Another quantitative value that is measured is delays that accumulate in the system. Here both the total delay should be measured and also the respective delays at each station. From this information one can also assess the queues that build up at each location.

Two more factors must be assessed, which are important for the decision making. The first is the reaction time of the station, that is, the duration between the time the information is presented and the time the decision is made (this does not take into account any delays before it is presented at the station). The second is the scale of priorities used in the system and its effect on the queues and the delays.

The qualitative analysis of the information must start from the source, giving the number of messages originating there as a function of time. A separate analysis is then performed on each path and for each intermediate location. At the decision location, another qualitative investigation is conducted to define the type and function of the decision made and how this affects the transformation of the information.

The complete qualitative to quantitative evaluation of the current system is then returned to the management for approval. No further design can be conducted without the assurance of the management that the designer fully understands the performance of the current system.

When the evaluation of the current system is available, it should be examined again, using operational research techniques. First, an analysis is conducted of the current system to record its advantages and disadvantages. The advantages indicate the predominant operations which do not require changing, while the disadvantages indicate the problems and the shortcomings of the current system. From this analysis the new system can be developed presenting the skeleton of the most effective operation. (During the operations research study there might be a need to return to the management for further clarification or for changes in the fundamental postulations. The latter could arise if the postulations prove to be either not effective or not feasible.)

The operations research lays the foundations for the new system and its plan could be formed by adding to it the growth in quality and quantity that is expected. New operations, speeds and facilities are introduced into the new system and these must be taken into account in the design programme. Although the quantitative growth is a function of the organization, it is not an easy figure to measure, and a lot of intuitive guesswork is entailed. The new system completely revolutionizes the operation, with new facilities and a far higher volume of information transfer. More so, once the new system is in operation, there may be a further call for new requirements which are currently unforeseen. Psychologically, once the system is in operation, the user realizes the significance of the new power, and this encourages and impels him to demand more facilities and effectiveness. The designer should try to envisage all possible requirements that will affect the volume and he is advised to allow for a good safety margin.

The programme proposed by the operations researchers is a system based on a theoretical analysis of the problems. It does not necessarily consider the further points such as cost and human performance. A model of the system should be constructed so that it could be simulated with the aid of a computer. All the variables that could affect the system operation must be considered. The result of the simulation analysis could show that the system is not feasible and requires a change in the fundamental postulations. It could also show that the facilities and/or the quantities selected are not practical, necessitating its return to the designer for modification of the requirements. The designer must compromise between the conflicting requirements, and the amended programme must then be re-simulated.

The simulation investigation should consider three topics, viz. the best system performance, the best equipment to be used and the best utilization of man-machine interaction. The last item could be carried out by constructing a small model of the system with actual displays fed by a computer which simulates the information required for the decision making. This simulation can analyse the reaction time of the human operator.

Simulation is considered to be one of the regular tools used by operations research; in spite of this, it is discussed here separately to emphasize the need for simulation.

Once a compromise has been reached and the system values of the facilities defined, the master plan of the new system can be constructed. The master plan is the programme which specifies all the details for designing the complete system. It gives the general specification for the whole system design, the multi-year budget and the policy for implementation. The latter includes the priorities in developing the models and subsystems according to the general programme of the organization. The master plan specifies the subsystem segmentation, giving the operational programme for each one. The scheduling of the subsystems is determined by the short-, medium- and long-range transaction plan specified by the management plan.

There are two distinct approaches that can be adopted by the designer when scheduling the transactions of the subsystem.

(a) An exhaustive research study, analysing each module of the whole system and defining in detail each element that could be used. Such an investigation could take a number of years and by the time it is completed might be out of date. There is also the danger in this approach that the research turns out to be the target of the work rather than the system design. The results of this work could be merely piles of books which, however impressive, do little to advance the project.

(b) Immediate purchase of equipment so as to demonstrate quick results. People enjoy seeing flashing lights and are impressed by working equipment. Some designers introduce equipment from the initial stages so as to gain the sanction of the customer. Other designers purchase equipment so that they can simulate the new system operation with the field operators. This is a most dangerous approach, since it is hard to simulate the whole operation with various bits and pieces of equipment. It must be stressed that any failure in the initial stage could cause a change in management policy and the discontinuance of the project.

Of the two approaches the writer is more in favour of the first, but for the practical reasons stated the best design approach must be a compromise between both suggestions. A research study of the system is essential and no purchasing should be allowed before the skeleton of the whole new system has been formulated. Only after the basic master plan and the integration specifications are available can the first subsystem be designed. Then one team can concentrate on the subsystem design while the main team continues the research of the complete system. The first subsystem should be small enough to be implemented rapidly. Its purpose is only to demonstrate the potent capabilities embedded in Communication Command and Control systems. This first subsystem is destined to rouse the interest of the customers and induce them to commit themselves to support the proceeding stages.

2.6 INTERACTION BETWEEN THE SUBSYSTEMS

The Command and Control system has a most complex structure and may comprise a large number of subsystems. Each of these may perform a separate operation and, accordingly, the development of each subsystem could be regarded as an independent project. Furthermore, the development of each subsystem may be concluded at different times with possible gaps of years between them. Nevertheless, all the subsystems are part of a large project aiming at the same goal. Many elements of the complete system may be shared between a number of all the subsystems; some may share all their facilities with others, while others may only share one or two of their facilities. There are at least three main facilities that are shared by most of the subsystems.

(a) *Communication*

The various subsystems may use the same communication lines, and so, they may also use the same line equipment and message-switching centres. The interaction requirements here are speed, message formats, mode of transmission, etc.

(b) *Processing*

The same computer or even the same files could be used for a number of subsystems. The interaction requirements here will be the use of the same computer language, the same communication procedure and even possibly the use of the same programme.

(c) *Display*

While the information may be collected and processed by different subsystems, the may have to be displayed at a centre using the same display for presenting the information and for decision making. The interaction requirements here are complicated, as a whole message format and the communication procedures must be the same.

The Command and Control system, being a gigantic project, may require the whole system to be developed in a modular form with each module added when necessity arises. That is, not only are the subsystems but also the modules developed at different times. The segmentation into modules and subsystems creates a danger, as the system is liable, if not properly designed, to look like an overlay of patches. To overcome this problem, strict interaction rules must be enforced, so that all the subsystems will be assembled and operated together (see Fig. 2.13). It is not important if each subsystem employs different techniques (such as those that result from using different data transmission speeds) provided they are considered during the initial design stages and provision for acceptance is inserted into the system. The interaction requirements must be specified before any

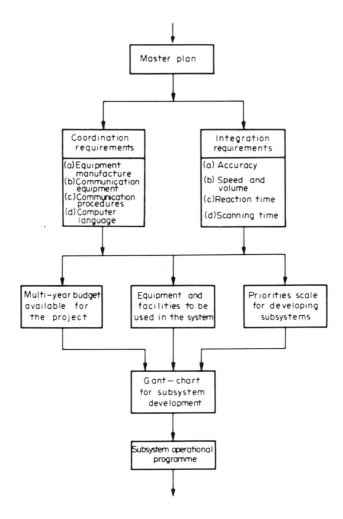

Fig. 2.13. Subsystem interaction specification.

items may have to be displayed at a centre using the same display for presenting the
the subsystems.

The reader may have observed that in all the preceding sections, the term integration
was used, while in this section the term interaction is adopted. The accepted term is
indeed 'integration', but the writer feels that a broader term is required. The term
'interaction' signifies that the problem is not only the integration between the various
subsystems when they are assembled together but also the coordination between the
subsystems when regarded as separate units. (It must be remembered that each subsystem
may operate separately prior to integration.) To cover both integration and coordination
requirements, the author has introduced the term 'interaction' requirements, in the hope
that it will be generally accepted.

The integration requirements presumes that there may be differences between the
various subsystems, while it acknowledges the requirement that all the subsystems must
be operated together. Special elements must be introduced in each subsystem which will
enable them to communicate and operate together at a later stage.

Some of the items which must be considered are as follows:
(a) The accuracy required from each subsystem.
(b) The speed and volume of the data originating from each subsystem.
(c) The scanning (or sampling) time of each terminal (or sensor).
(d) Speed of reaction from each subsystem.
(e) Services provided by each subsystem.
(f) Geographical location of the subsystems for sharing communication lines.
(g) Processing requirements for sharing computer facilities.
(h) Volume of data for sharing display facilities.

As the subject of the subsystems and modules is regarded as part of a single complex
system, means must be provided for the future amalgamation of all the subsystems and
modules in one vast configuration, as defined by the integration requirements. In view of
the differences between the various subsystems and modules efficient coordination within
the system can only be achieved if (a) the design method is strictly uniform and (b) if all
the subsystems and modules are constructed with identical elements.

In Command and Control systems the coordination requirements are extremely
important as they usually contain many subsystems each dealing with different topics.
The fact that each subsystem may be developed at different times makes the problem
more crucial, as the engineering techniques may have changed in the meantime.

Some of the coordination requirements are presented here:
(a) Defining the standardized equipment to be used in all the subsystems and/or the
     manufacturers to be used throughout. As the system as a whole will be
     maintained by the same service engineers, all the equipment should be identical.
(b) Defining the same communication equipment that must be applied to all
     subsystems. The subsystem may be miles apart but they may still have to
     communicate with one another or merely use the same communication lines.
(c) Defining the communication procedures so that all the subsystems can
     communicate with each other. This should include transmission codes,
     synchronization methods, block size, message formats, error correction codes,
     etc.
(d) Defining the computer processing languages so that all the computers from the
     various subsystems can converse with each other. Furthermore, the subsystems

may be required to share their resources and even their data files.

Part of the interaction specification also includes the time-table for the development of each model of the subsystem. This is based on the management priority scale for organization growth requirements which subsequently produces the priority scale for the development of the subsystems. The timetable must be defined as accurately as possible, but it still has to remain flexible to allow for change when the actual development plan is designed. It is advisable to use a gant-chart for this design stage and a 'pert' for the final stage.

2.7 SUBSYSTEM TECHNICAL SPECIFICATIONS

The master plan provides only a general description of the system design, but not the details. These are given in the technical specifications which are produced for each subsystem respectively. Here, the specific equipment is identified, and likewise the subsystem construction and equipment interconnection. Also included are complete details of the computer programmes and information details to display. The technical specification is the final stage in the design of each unit, and the next stage is that of the actual development.

The master plan presents the operational programme for each subsystem, and describes the objectives and configuration of each of them. The objectives define the system performance of the subsystem, that is, the operation, services and facilities expected from them, while the configuration defines the complete geographical formation of the whole subsystem, that is, the location of each terminal and centre, including the volume of information transferred between them. The volume is calculated by the analysis of the volume measured in the current system, the expected growth and a safety factor.

The objectives and the configuration of the subsystem are the design targets, which must then be achieved by the design execution: the targets are permanent and confirmed while the means for achieving the respective targets are flexible and floating. The design details are a matter of compromise between the conflicting design requirements, as shown in Fig. 2.14.

The design transaction has to specify the subsystem prospectus by producing diagrams, plans and part lists. The design prospectus may be divided into three related studies:

(a) *Quality and quantity*

The volume to be handled by each unit, the accuracy demanded of each operation, the subsystem efficiency and the priorities for information messages. It also includes the reliability required from the whole subsystem and from each unit according to their priorities. The redundancies within the subsystem to enable the required efficiency and reliability to be achieved.

(b) *Performance*

The operations the system is to perform in the collection of data, processing and decision making. It specifies who performs the operation, man or machine, how the operation is performed, and where and when it is performed.

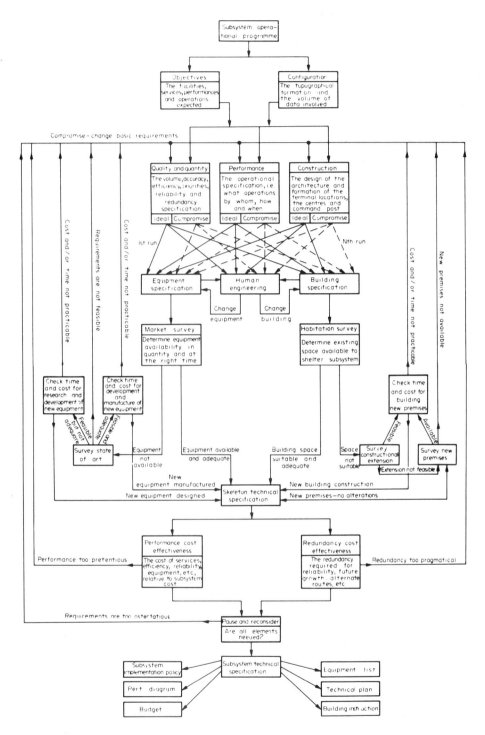

Fig. 2.14. The design procedure defining the technical specification.

(c) *Construction*

This is the design plan which specifies the architecture and formation of the subsystem. It specifies the equipment used and the structure of the terminal location, the processing centres and the command post.

The jigsaw character of the three design studies derives from the conflicting operational procedures. The method adopted is one of trial and error. First an ideal system is formed with full basic details. This is then modified in the light of the approved details of equipment and building specifications and the human engineering requirements, and a second run is performed. A number of design runs will presumably be needed before the final system is established.

2.8 INTRODUCING THE NEW DEVELOPED SYSTEM

After the system has been designed and developed, there is still the problem of introducing the new system to the operators. This is by no means a simple problem. Many Command and Control systems have been designed successfully, but because of inefficient introduction have proved a failure in operation. As the system must be operated by humans, it is vital that they know how to manipulate it efficiently. Hence means must be found of introducing the human operator into the system, as an integral element on a par with the other system elements.

Before dealing with the actual system presentation, an analysis of the problems that will confront the designer is warranted. The designer should anticipate possible opposition and disappointment in the new system due simply to human characteristic qualities.

(a) People may be scared that the new system (i.e. the computer) will dominate their life.

(b) People may be inherently frightened of any changes in their routine life.

(c) People may be conservative and suspicious of anything new or modern.

(d) People are desperately frightened in case the new system will cause redundancy, leading to their dismissal. Even when this scare is generally unfounded, they will still remain restless with regard to the new system. In Command and Control systems this fear could also apply to the management level, since the system is intended to help in decision making.

(e) The operators may not understand the system logistics, although they know what problems the system is intended to solve. They are liable to fail to follow the specific logic of a new system presented to them which is very different from the one they are used to.

(f) The operators will not "fall in love" with the new system, as they may remember the 'quiet' routine work they had in the old system. The new system will require rapid action from them, as the machine cannot wait. The system being automatic will not allow any elements, i.e. humans, to be tardy. Furthermore, in the old system the operators could neglect or forget to deal with items and no one would know about it, whilst in the new system everything is recorded.

(g) The operators may not take full advantage of the new system capabilities. Even after the operators have been introduced to the system, they may not fully appreciate the facilities which could be exploited. This is a well known-problem in the computer field, where a large portion of the C.P.U. is used only for short periods. In Command and Control, which is a more complex system, the problem could be even more critical.

(h) Even if the system answers all the basic requirements, there could still be criticism that the system does not perform what the operators hoped it should do. People do not appreciate that the requirements are not a stable issue. Once the system is in operation and they realize what it can do, new requirements, unthought of before, arise. The operators are then surprised that such a sophisticated system is not capable of solving their new requirements on the spot.

(i) Again, even if all the problems raised above have been solved, there could still be some disappointment in the new system. There is simply no such thing as an ideal system, as there will always be a call for additions and expansions. The initial design must consider the possibilities of these contingencies and insert flexible means for permitting the further introduction of new or different services.

(j) In spite of the thorough design study, the anticipated loads may be considerably different from those encountered in practice. This also applies at the interaction point of man–machine, where the operator's reaction cannot be predicted until the system is actually in operation. In order to anticipate this problem, monitors must be inserted into the system, both in the hardware and the software, so that the statistical performance of the system is continuously supervised. This will enable the changing of the system configuration as the load factor increases above the predicted measure.

(k) A system is always in danger of being out of date by the time it is implemented, since the state of the art, of the equipment and techniques used, may have advanced in the meantime. The designer must therefore not aim at designing the most advanced system but rather a system that can operate at a given date. That is to say, the designer must fix a demarcation date, beyond which no more changes are introduced, otherwise the system would never get completed.

Some of the points mentioned here have to do with natural human failings, which the designer cannot of course change. He can, however, reduce the extent of the impact of introducing the new system by making allowances for them.

(a) Whenever possible, during the development stages, the designer should demonstrate the system capabilities by means of a series of lectures.

(b) The new system should be introduced in stages and not with a sudden revolutionary change. Only after the operators have fully grasped the control of the specific stage can the second stage be introduced.

(c) The operators should undergo a detailed course before they start operating the system. The system should be taught by a model simulator, controlled by a computer which can simulate all the system performances.

(d) The operators should be allowed a free play with the system (or preferably the simulator) by operating any of the knobs they wish. In this way the operator can learn the extent of the system's power. By this simple play the operator may be relieved of some of the fear he may have had and also satisfy his normal curiosity. This of course can be practical only if it was foreseen in the design stages. Many systems fail if random knobs are pressed.

(e) A short introductory course should be presented to all the personnel who may be assisted by the system, that is, apart from the operators.

(f)   A second refined course for the operators which is presented after the system has been in operation. The operators by now have learnt the system skills, but require a second course to increase their competence, and to show them means of increasing the efficiency of the system by combining some of the operations.

2.9 REFERENCES

1.   Hamsher, D.H. (Editor-in-Chief), *Communication System Engineering Handbook*, Chapter 1 by Roy K. Andres, "System design requirements", McGraw Hill Book Company, 1967.
2.   Martino, R.L., *MIS-Methodology*, MDI Publications, Wayne Pennsylvania, 1969.
3.   Bocchino, W.A., *Management Information Systems – Tools and Techniques*, Prentice-Hall Inc., 1972.
4.   Birmingham, H.P., "The optimization of the man–machine control systems", *1958, IRE WESCON Convention Record*, Pt. 4, pp. 272-276.
5.   Marzocco, F.N., "The human factors laboratory as a system design tool", *1960, IRE WESCON Convention Record*, Pt. 4, pp. 130-133.
6.   Freed, A.M., "Measuring human interaction in man–machine systems", *1960, IRE WESCON Convention Record*, Pt. 4, pp. 189-201.
7.   Schneider. R.H., "Human factors in the establishment of system design requirements", *1960, IRE WESCON Convention Record*, Pt. 4, pp. 127-129.

CHAPTER 3

Sensor Base Data Collection

3.1 INTRODUCTION

In order to provide the organization with better efficient control of its resources a constant feedback of information is required from the lower to the upper levels of the closed loop Command and Control system. This is achieved by means of sensors which collect data from various positions in the system for transmission to the upper levels.

In most Command and Control systems, the information that is inserted into the system is already in digital form. That is, the data is generally inserted by alpha numerical terminals which are converted into binary coded representation. There are some Command and Control applications, however, where the data inserted into the system is obtained from sensitive sensors in analogue form and subsequently converted into digital representation.

The subject dealing with analogue measurements and their transmission is known as Telemetry. This chapter gives an elementary introduction to telemetry systems, so that the reader can get a general idea of the subject to assist him if he may require it for the design of Command and Control systems.

Telemetry, by definition, is the collecting of data from inaccessible locations and transmitting the information to accessible positions. It is in general the process of measuring a quantity collected from remote sensors which is to be stored, displayed or used to actuate a process. In some Command and Control systems, telemetry methods are used for the data collection, and other methods are used for processing it and correlating it with other data. The data collected by the telemetry methods must be transformed into meaningful information for possible decision making. In other words, all the analogue information must be translated into digital form.

Although telemetry deals with analogue information and Command and Control with digital information, they must be regarded as a single system. Where telemetry is used for the data collection of the Command and Control system, it must be designed as an integral part of the system and not as a separate scheme. By regarding the telemetry as one of the techniques and not as two separate system schemes, its whole design aspect is changed with the following advantages:

(a) Increase in flexibility.
(b) Optimum usage of the communication network.
(c) Centralized control operation.
(d) Simplified control system.
(e) Storage and retrieval of past measurements.
(f) Easy correlation with past history.
(g) Means for processing the collected data to a meaningful information status.

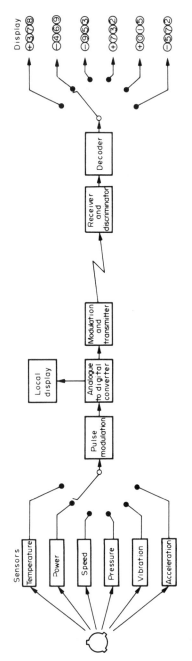

Fig. 3.1. Typical s mplified telemetry system.

(h)   Transferring the operations of processing and displaying, from telemetry to other sections of Command and Control.

The information that flows in telemetry systems is in a single direction, whereas a Command and Control system operation is a closed-loop system with the telemetry as a possible node of the loop. In Command and Control, the information which has been collected by the telemetry system is used for decision making and for real-time control of the decision implementation.

3.2 TELEMETRY SYSTEMS

A telemetry system is intended to measure the status operation of a number of remote locations and to transfer this status monitoring reading to a centralized installation. One of the main objects of any telemetry system is to get the data collected into a meaningful and usable form. A typical simplified telemetry system is shown in Fig. 3.1.

The installation under control is continuously monitored, to check its operational status. This is performed by a series of sensitive sensors which perpetually measure the system operation. The sensor output readings are scanned at fixed intervals, producing a train series of pulses with different amplitudes, with each pulse amplitude representing a distinct measurement. The scanning operation uses time division multiplex techniques, where each reading receives a different time slot in a sequential pulse train. The various samples of amplitude pulses obtained by the multiplex scanning is already in the form of what is termed pulse amplitude modulation.

The random amplitude level of pulses may be further modulated so as to bring it to a form less susceptible to noises, as will be explained later. For most Command and Control applications the analogue readings must be translated into digital form. The analogue to digital conversion of the amplitude level of train pulses produces a sequence of binary pulses, where each set in the sequence represents a specific instantaneous sensor reading.

Before the digital information can be transmitted, it must be modulated once again back to analogue form so that it will not be distorted in the communication network. (The transmission aspect of the digital data is discussed in greater detail in the following chapter.)

At the receiver the information is demodulated, returning it back to digital form. The information received is in serial form, each set of the sequence belonging to a different monitoring status reading. The readings must be distributed to the correct numerical displays, using demultiplexing techniques.

Telemetry programmes could assist in Command and Control systems by using one or all of the following types of monitoring status.

(a)   As a warning in the case of a failure. This only requires the measuring of the extreme values and reporting the data in ON–OFF form.

(b)   Continuous reporting of the installation status operation.

(c)   Reporting only changes in the installation status operation.

In some telemetry publications, a number of technical terms are used in a context which may be misleading when applied in Command and Control. To prevent any misunderstanding, the terms are clarified here.

(a)   Pulse amplitude modulation (PAM) results by sequentially sampling a waveform at fixed regular time intervals. That is, the carrier is modulated with pulses whose amplitude level carries the information.

(b) Time division multiplex (TDM) is the concept of sharing the time between a number of channels. That is, the information in each channel is sampled in a regular sequence. PAM uses TDM techniques for scanning the waveforms. It is a mistake, however, to regard all time division multiplexors as pulse amplitude modulation. The time period allocated for each sample in TDM may consist of a series of pulses and not just a single amplitude as in PAM.

(c) Analogue to digital convertors (A to D) is the digital translation of an instantaneous amplitude level.

(d) Pulse code modulation (PCM) refers to the digital representation of a waveform which is analysed both by time and amplitude samples. PCM usually deals with many waveforms, where each set in the pulse train is associated with a specific time and level of a different waveform. Pulse code modulation is closely allied both with TDM and with A to D by using these techniques. Here, it is a mistake to call a PCM system an A to D system or a TDM system.

3.3 PULSE MODULATION

The measurement waveform obtained from the sensor output must be conditioned by a suitable conversion, to be compatible with transmission requirements. The waveform must be sampled, or in other words, be pulse modulated. Telemetry schemes use a variety of pulse modulation techniques, as shown in Fig. 3.2. The modulation technique selected depends on a number of factors: (a) why the measurement was required; (b) where it is to be transmitted; (c) how it is to be displayed; (d) in what form it will be displayed.

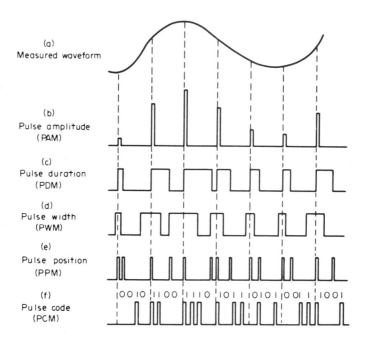

Fig. 3.2. Pulse modulation.

Pulse amplitude modulation (PAM)

The basic modulation technique is that of pulse amplitude modulation. The waveform obtained from the sensor output is scanned at fixed intervals. This presents a train of instantaneous amplitude levels, appearing at fixed intervals and representing samples of the original waveform.

The original waveform can easily be reproduced by passing the pulses through a low-pass filter. The most popular method is to use an operational amplifier with a capacity feed-back.

The distance between the sampling pulses must be rapid enough so that no information is lost between sampling instances. According to Shannon's theorem the sampling rate should be at least twice the highest frequency contained in the original waveform. (This theorem states that if a function of time contains no frequency higher than f, it could be completely determined by a series of points spaced $\frac{1}{2}f$ apart.) In fact, in many telemetry schemes a much higher sampling rate is used in order to preserve all of the original information. The rate, however, is much slower than the speeds that flow in the Command and Control systems, and so this rate presents no handling problem.

Pulse amplitude modulation is the simplest modulation technique. It is easily produced by using a two-input AND gate, where one of the inputs is the waveform and the other is a train of clock pulses.

The main drawback of pulse amplitude modulation is that it is very susceptible to noisy lines. Any noise can distort the amplitude pulse level and thus deform the whole meaning of the reading.

Pulse duration modulation (PDM)

To overcome the amplitude-level noise distortion, the amplitude pulses are transformed to new pulses of fixed height but with variable duration. The noisy lines cannot affect the pulse level, as all the pulses are of identical uniform balance. The sensor measurement information is thus contained in the pulse duration rather than the pulse height.

This technique is known as pulse duration modulation, where all the pulses start at the same relative point, controlled by the clock timing. The information is retained in the position of the trailing edge of the pulse. In other words, the information is held by the area bounded by the pulse in the same way as in pulse amplitude modulation. Hence the original pulse amplitude level could easily be reconstructed by charging a capacitor with the variable duration pulse.

Pulse duration technique is far less susceptible to noise than the pulse amplitude technique, since most of the disturbances affect the pulse height rather than its duration. Nevertheless, any distortion of the pulse shape will be misrepresented by the reconstructed information reading. This distortion could easily come about by the inductance and/or the capacitance of a transmission line. As the sampling frequencies are increased, this problem could become more acute.

Pulse duration modulation is used in some magnetic tape recording processing. The advantage of PDM over other recording processes is the ability to record many signal channels of information.

Pulse width modulation (PWM)

Pulse width modulation is similar to pulse duration modulation except that here both the leading and trailing edges vary with the information context. This modulation technique may be regarded as duplication pulse duration, as both the leading and trailing edges of the pulse are varied in the same portion relative to the clock position. Although theoretically it has an advantage over pulse duration, it still has the same drawback of being susceptible to pulse shape distortion.

Pulse position modulation (PPM)

Another solution to the distortion problem of the pulse shape is to transmit only the two edges of the pulse duration. This modulation technique is known as pulse position modulation and is based on the pulse duration technique, but here pulses of fixed amplitude and duration are transmitted for each time sample instead of a variable pulse width.

Two pulses are transmitted for each time measurement sample, the first coinciding with the leading edge and the second with the trailing edge of the pulse duration pulse. Since all the pulses in PDM start at the same time, i.e. to coincide with the clock timing, the leading pulse of PPM is synchronized with the clock pulse. In consequence, there are some telemetry systems where only the trailing-edge pulses are transmitted.

The original pulse duration could be easily reproduced by simply feeding the pulse train sequence into a T flip-flop.

The pulse position modulation is superior to all the three modulation techniques described above, as it is not susceptible to any amplitude or shape distortion. However, spikes in a noisy line could be picked up, amplified and shaped by the system, consequently deforming the information, since they could be mistakenly acknowledged as real pulses. (This is a common problem in data transmission, and will be discussed later.) Pulse position modulation also suffers from the requirement of a wider frequency bandwidth for transmitting the information, due to the narrowness of the pulse. This is a standard problem of digital pulse transmission, although here the problem is more critical, as only single pulses are transmitted and their position must be accurately defined.

Pulse code modulation (PCM)

For Command and Control applications the measurements must be transformed into digital form. The methods of transforming the analogue pulses into digital coded information is known as pulse code modulation.

The original sensor output waveform is not only vertically sampled but also horizontally quantized and the result is presented in a train of binary pulses. The vertical samples are performed by a clock scan and the horizontal by a level scan.

The pulse code modulation represents the most efficient and the most practical coding technique. In view of the importance of this technique to Command and Control systems it will be discussed in greater detail in Section 3.5.

PCM also requires a wide bandwidth for transmission of the pulses, as in PPM, but their position is not so critical here, as the pulses all coincide with clock pulses, while in PPM the pulse position varies with the information context.

3.4 MULTIPLEXING OF ANALOGUE WAVES

The whole telemetry programme is based on the reading of a number of measurements from remote locations and transmitting samples of these readings to a centre installation. The measurement readings generally vary slowly and the sampling information produced is relatively small, compared with the possible transmission speeds. A separate transmission channel for each measurement (as shown in Fig. 3.3 (a)), proves to be most inefficient and wasteful. The alternative solution is to use a common transmission channel (as shown in Fig. 3.3 (b)) where several measurement readings are transmitted simultaneously. The technique of transmitting a number of channels over a single telecommunication link is referred to as 'multiplexing'. With this technique a number of signals are combined together so that they may be transmitted as a single signal. The original signal may be separated again by a method referred to as 'demultiplexing'.

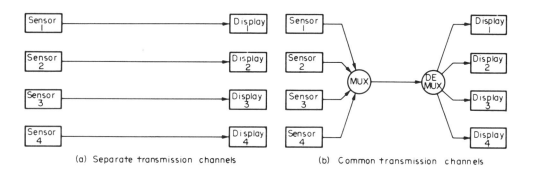

(a) Separate transmission channels   (b) Common transmission channels

Fig. 3.3. Principle of multiplexing.

There are two basic methods of multiplexing techniques: frequency-division multiplexing and time-division multiplexing. In both techniques a number of different information measurements may be transmitted in parallel without substantially degrading the accuracy of the information.

Frequency-division Multiplexing (FDM)

In frequency multiplexing the transmission bandwidth is divided between the various sensors, as seen in Fig. 3.4. Each measurement reading is allotted to a separate frequency bandwidth within a common wide bandwidth of the transmission path.

The measurement readings at the output of the sensor are fed through a low-pass filter in order to restrict the bandwidth required for the transmission. The upper cut-off frequency of the filter is selected by the highest frequency component of the measurement reading, which must be kept as high as possible so that the reading may be transmitted without substantially degrading the information, and must be kept as low as possible to allow for many sensor output readings to utilize the common transmission frequency bandwidth.

Each measurement is then modulated on a separate carrier signal. All the measurements are spaced apart by a fixed frequency, with each modulated measurement

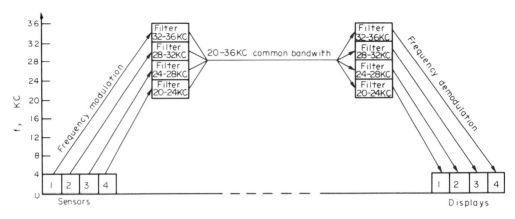

Fig. 3.4. Frequency division multiplexing.

restricted to its own bandwidth. Between each band and the next there must be a guard band to ensure that the modulated signals do not spill over into bands of other modulated signals. The modulated process may be amplitude, frequency or phase modulation (as will be discussed in the next chapter). Before the modulated signal is transmitted to the common link it must be fed into a bandpass filter to guard against any upper or lower frequencies creeping into the adjacent channels.

At the receiving end a reverse process takes place. The common link feeds a parallel set of bandpass filters. The separation of the channels is easily accomplished by these filters provided they were properly restricted at the transmitter end. Each channel separately is demodulated, thereby presenting the reconstructed measurement reading.

To achieve efficient use of the common transmission bandwidth, each measurement should not be allocated with an equal frequency range but only with the actual bandwidth it requires. This would enable the same common bandwidth to be shared between a larger number of sensors, an arrangement which would not be possible if the bandwidth was shared equally.

Time-division Multiplexing (TDM)

Time-division multiplexing is similar to pulse amplitude modulation already discussed, although here the sampling is performed on a number of measurement readings, scanned in turn, as seen in Fig. 3.5. As in frequency division, all the measurement readings use the same common transmission link. In time-division multiplexing all the channels use the same portion of the frequency spectrum, but not simultaneously. The time spectrum is shared equally between the different channels.

All the sensor outputs are continuously scanned at fixed intervals. This process produces a train of pulses of various amplitude levels. Each sensor reading receives a time slot within the time pulse sequence. After completing a set of all the sensors it starts once again with the first sample. The pulse train of variable amplitudes is transmitted into the common link, each sample interleaved with another sensor's sample.

The technique of time multiplexing is completely different from frequency multiplexing. In frequency multiplexing a continuous measurement reading is

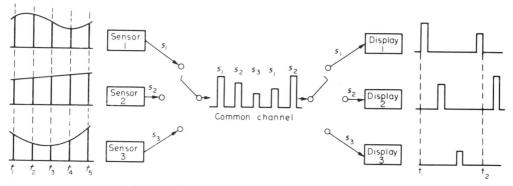

Fig. 3.5. Time-division multiplex—principle operation

transmitted, while in time multiplexing only discrete samples of particular instants are transmitted. Frequency multiplexing is used when the original waveforms are to be transmitted, whereas time division multiplexing is used for transmitting samples of the original waveform. In other words, in time multiplexing the information is in pulse form and is applied especially for the transmission of digital information. For telemetry schemes used in Command and Control systems, time multiplexing is the more important of the two types of multiplexing.

Frequency- and time-division multiplexing have a much broader application than that discussed for the telemetry programme, and this will be discussed in greater detail in Chapter 5.

For simple telemetry applications the time slot allocation for each measurement reading is a single pulse, as shown in pulse amplitude modulation. However, each time slot could be extended to contain a set of pulses as shown for pulse code modulation.

The process of achieving time-division multiplex is very simple (as shown in Fig. 3.6). Each sensor output feeds a separate two-input AND gate. The other input to the gate is fed from a series of clocked sequential pulses, which are obtained from decoding a counter. The amplitude of the clock pulses must be much higher than the largest sensor output. All the AND gates feed an OR gate, this producing the required sequence of variable amplitude pulses.

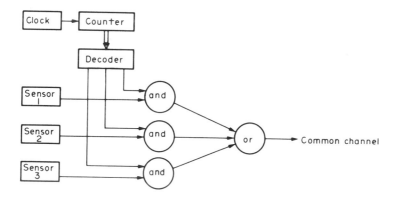

Fig. 3.6. Time-division multiplex—logical diagram.

At the receiving station the exact measurement samples must be reproduced so that the information can be displayed. The programme is based on precise timing of the time slots, both at the transmitter and the receiver ends. Thus it is most important in time-division multiplex to synchronize both ends. This is usually done by sending a synchronization code at the end of each complete scanning cycle.

3.5 PULSE CODE MODULATION (PCM)

Pulse Code Modulation is a technique where all the analogue data is coded in binary notation. The train of PCM pulses is already in a format which could easily be administered by the communication switching network and also be transacted by the computer. The sequential pulse train of PCM could also contain further information, to enable easy manipulation, such as synchronization pulses and sign codes.

For all Command and Control applications the information which flows in the system must be in computer language, that is, first in digital form and then in a specific format. The object of Command and Control is not only to transfer data from remote locations to a central installation for it to be displayed, but also mainly to permit processing of the various data readings into meaningful information for decision making. PCM is the modulation technique to be used in applications where the telemetry data is to be processed by a computer. For display purposes only, the other modulation techniques too could be feasible, but if any logical manipulation is required it must be in digital form.

Pulse code modulation is an ideal technique for the transmission of analogue data, as the data is freed from attenuation and phase distortion. It is also not susceptible to any variation of the local changes when transmitted over a long distance. The mere presence or absence of a pulse can be easily recognized when the distortion is present. Furthermore, spikes which may be amplified as pulses will not be recognized as data if these pulses do not appear at the right sequential timing. In other words, pulse code modulation is the most efficient of the practical coding and modulation techniques.

Pulse code modulation has been known since 1937, although its development was halted for many years owing to the complex nature of the circuitry involved. It is only in the past few years that its potentiality has been appreciated, and it is now playing a most important part in the digital handling of analogue waveforms. All the modern fully electronic telephone exchanges use PCM techniques for the coding of the speech transmission and for the control of the exchange. Pulse code modulation development had to wait for the introduction of miniature integrated circuits for analogue and digital applications.

In pulse code modulation, the sensor output waveform is sampled periodically as in PAM and then the amplitude values are represented by a coded sequence of several pulses of equal amplitude and duration. When considering individual sensor outputs three successive operations are needed to transform the analogue waveform into a series of digital pulses (as shown in Fig. 3.7 (a)). Each individual waveform is sampled at a rate twice that of the highest frequency (as in PAM technique already discussed). Then each individual sample amplitude level is compared with a standard scale of fixed references. The closest value of the comparison is coded into binary form (as will be explained later). The train of pulses obtained is in a state where it can be transmitted to a local centre,

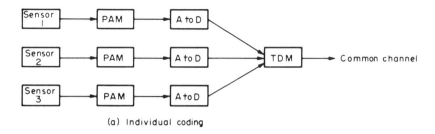

(a) Individual coding

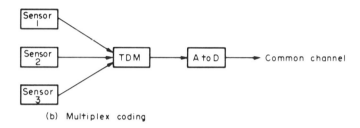

(b) Multiplex coding

Fig. 3.7. Basic pulse code modulation telemetry schemes.

where it is then time multiplexed with other sequences of coded pulses. The multiplexed information can then be transmitted on a common line to the centre installation. The multiplexing may be done by taking each bit separately or, more practically, taking a complete word from each measurement.

When a number of sensors are situated in the same location it is far more efficient to perform the multiplex operation near the sensors (as shown in Fig. 3.7 (b)), thus avoiding the double operation and the need for separate analogue to digital converters for each sensor. Although the first method seems wasteful, there are some applications where it is unavoidable. In the case where a single A to D is used, the output signals from all the sensors are sampled sequentially, as described for the time-division multiplexing. This presents a series of variable amplitude levels which are then converted to a series of digital pulses. The train of pulses consists of sets of sequential digital pulses, each associated with a different sensor instantaneous measurement reading (as shown in Fig. 3.8). Before each set of pulses, a synchronization pulse is usually inserted to indicate the beginning of the word. After a sequence of all the coded readings it returns to the first sensor for the second sampling reading.

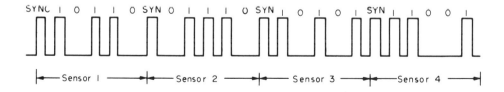

Fig. 3.8. PCM pulse train.

The advantages of pulse code modulation are summarized below:

(a) PCM data is already in computer language and thus can be fed directly into the computer.

(b) PCM data is easily manipulated for any processing requirements.

(c) PCM data can be easily buffered for later use.

(d) The digital information transmitted is not directly proportional to the original analogue signal; hence it is much less susceptible to line noises.

(e) Changes and distortions in the digital information do not seriously affect the reproduced information, since it could easily be corrected by the filters.

(f) PCM can efficiently utilize the frequency bandwidth (relative to FDM).

Nevertheless, pulse code modulation does suffer from disadvantages which have some bearing on the design of these schemes.

(a) The original analogue waveform will never be reproduced exactly, because of the quantizing errors.

(b) Requires a relatively wide transmission bandwidth.

(c) Requires an elaborate synchronization programme to distinguish between the various sequential channels.

(d) The PCM data needs to be modulated or requires to be reproduced and reshaped if it is to be transmitted over long distances.

3.6 ANALOGUE TO DIGITAL CODING CONVERSION

The basic feature of pulse code modulation is the conversion of the analogue waveform to digital coding. As already explained, the amplitude of the waveform is first sampled vertically at fixed time intervals, producing a set of variable amplitude levels. The pulse samples are then quantized horizontally at fixed levels (as shown in Fig. 3.9). Each

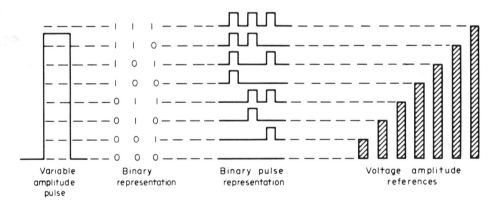

Fig. 3.9. Analogue-to-digital encoding using voltage references.

individual sample level is compared with a standard scale of fixed voltage references. Each measurement is assigned the lowest closest value of the comparison. This approximation rounding-off process is called quantization.

Each fixed voltage reference value is coded into a set of binary codes (as shown in Fig. 3.9). If 2^n discrete reference values are selected, then the binary code is constructed by n bits. The quantized uncertainty of the waveform is plus-minus half a sample spacing. Therefore, the uncertainty is associated with the least significant binary bit.

A code of 5-7 bits is usually used in most applications, depending on the problem involved. The number of sample levels is a direct function of the errors in re-creating the original waveform. In the example shown in Fig. 3.9 only eight samples are used, i.e. 3-bit code; it is obvious that increasing the sampling rate to 32 or to 64 or even to 128 will reduce the quantization error.

The binary code usually consists of an extra bit to signify the polarity sign of the sampled waveform. Although this may be a waste of one sample level (since the zero level is the same for both polarities) it simplifies the handling of the information for processing purposes. Thus a four binary code will contain only fifteen sampling levels, viz. zero and seven levels in each polarity direction.

The disparity between the original waveform and the quantized digital representation can be seen by the example given in Fig. 3.10. This shows the reconstructed shape of both the vertical and horizontal sampling. The pulse amplitudes shown are not those received from the time division multiplex but from the close approximation received from the comparison with fixed levels.

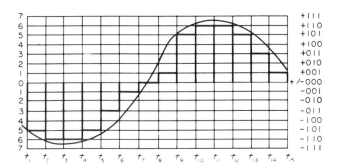

Fig. 3.10. Quantized approximation of sampled waveform.

Until now the description has assumed that the horizontal samples were taken by comparing them with fixed voltages having equal level spacing between them. This is on no account an essential requirement, and there are many schemes where the level spacing is in log form or any other variable spacing. In some applications the small weakest signals are important; hence more samples are taken near the zero level in order to distinguish the various weak signals. In other schemes the important measurement factor may be near the peak of the signal; when that is the case, the peak is sampled at less spacing than near the zero level.

As mentioned before, each set of binary codes has a synchronization pulse added to it. In many cases another pulse is also added, namely, the parity pulse required for error detection when the information is transmitted. This brings a binary code of n bits to a $n + 2$ bit code.

The example shown in Fig. 3.11 demonstrates the spacing of the bits in a PCM scheme of a telemetry programme. In the example in question the measurement changes at a rate

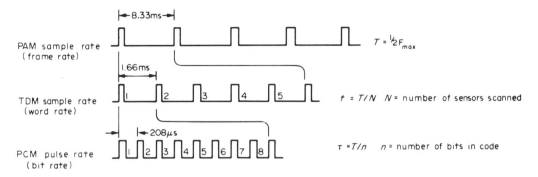

Fig. 3.11. Pulse rate in an example PCM telemetry programme

of 60 Hz, which is also the highest frequency. The minimum sample rate of each sensor output waveform is $T = \frac{1}{2}f\,max = 1/60 \times 2 = 8.33$ ms. This is a minimum spacing between the individual pulses in PAM of the sensor waveform. This is also the spacing of one complete scanning cycle, including a sample from each sensor in the location. The complete cycle is called a FRAME. In this example, the number of sensors that are to be scanned is five. Thus, the spacing of the pulses when sampling all the sensors is $t = T/N = 8.33 \times 10^{-3}/5 = 1.66$ ms. This is a result of a TDM which is called a word, and represents a code set of a single sample. This spacing must still be divided by the bit rate of the code to give the PCM pulse rate. If the horizontal sample rate in this example is 32 samples in each polarity direction, then a 5-bit code is required in each direction. To this code a sign bit must be added which brings it to a 6-bit information code. The PCM transmitted word must also consist of a synchronization pulse and a parity pulse, thus bringing the word size to 8 bits. The spacing of the bits in the PCM pulse train is equal to $\tau = t/n = 1.66 \times 10^{-3}/8 = 208$ μs. For this pulse rate, a transmission speed of $1/\tau = 1/208 \times 10^{-6} = 4800$ bits per second is required.

Analogue to digital conversion could be performed by a variation of schemes. Most systems use a comparator circuit, where the unknown voltage level is compared with a reference voltage and indicates whether the result is larger or smaller. Only one comparator need be used, but this requires a closed-loop feedback scheme. Figure 3.12

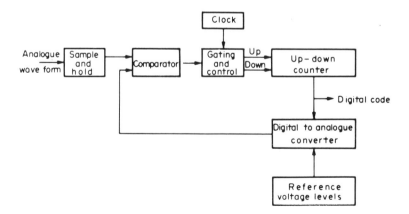

shows one of the solutions for the analogue to digital converter.

An Up–Down Counter is used for the binary code received from the A to D converter. The binary code is converted back to analogue form and the voltage reference received is compared with the input voltage sample. The counter is then shifted up or down, depending on whether the corresponding comparison is larger or smaller than the input. With this result, the counter is modified till it reaches the quantized approximate value of comparison. The counter is of n flip-flop stages, where n is the number of bits in the sampled word. The two extra pulses, for synchronization and parity, are added later.

3.7 SPACE-DIVISION MULTIPLEX (SDM)

Space-division multiplex is the accepted term for the establishment of the physical connection between terminals which are separated in space from the other connections in the network. It could be regarded as an extension of time-division multiplex. The design of space division multiplex involves the topology of the network and the choice of the crosspoint. Time division multiplex scans a number of inputs and multiplexes them into a common channel. Space division multiplex can do the same operation and perform a number of these operations in parallel.

Space-division multiplex is most popular in telephone circuit-switching, but it could easily be a suitable tool in telemetry for switching the measurement readings. Space switching could be most useful in Command and Control systems for many applications, such as communication line switching, multiplexing one large number of sensors into a limited number of parallel common lines, or even in the command post where a large number of displays receive their information from a number of sources. In all these possible applications the SDM acts as a communication controller. Its main advantage is that there is no limit to the number of terminals or to the bit rate.

Space-division multiplex is made in a modular form, where the load is distributed to ensure security of service when there is component failure. It enables the information to be transferred in parallel and in both directions simultaneously. SDM was originally designed to switch analogue information, but lately its utilization for digital application is increasing. Many of the modern telephone exchanges use SDM for switching PCM data. The idea of space-division multiplex is also used in computers, where the core store switch is one example.

The simplest space-division multiplex network is the cross-bar switch shown in Fig. 3.13. A cross-bar switch consists of horizontal and vertical bars which can be operated separately in such a way that they activate the relay-like cross-points attached to the bar. By suitable selection of the bars, a coordination cross-point is obtained, providing a path between one of the input rows and one of the output columns. A cross-point can operate only if it has been activated by the two perpendicular bars and is held locally.

Each cross-point intersection may consist of a number of switches in parallel. Thus, for each coordination selection, a number of switches are operated simultaneously. Figure 3.14 shows the basic selection network and its accepted representation. The number of input lines is represented by m, the number of output lines by g and the number of levels in parallel in each cross-point as p. The number of cross-points in the matrix is $m{\times}g$. In each matrix there could be g simultaneous paths within the matrix, provided $m \geqslant g$.

The initial cross-bars designed for the telephone industry consisted of relay network. In the last few years dry read relays have come into prominence as a standard switching

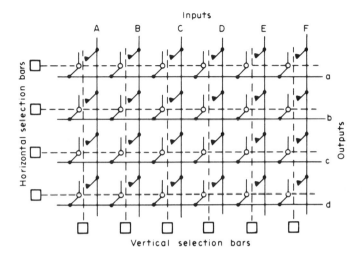

Fig. 3.13. The principle of the cross-bar switch.

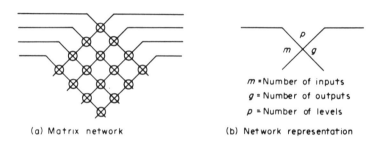

(a) Matrix network

(b) Network representation

m = Number of inputs
g = Number of outputs
p = Number of levels

Fig. 3.14. Space-division multiplex network switch.

element for the cross-bar. However, with the advance of MOS integrated circuits many fully electronic cross-bar switching networks have been introduced. A cross-point consisting of NAND gates can produce a three-dimensional space-division network. An example of a fully electronic SDM network for telemetry applications is shown in Fig. 3.15. The use of integrated circuits reduces cross-talk problems which are unavoidable in electromechanical switches. It has the advantage of simplifying the network and enables ease of expansion, besides being small, economical and above all reliable. The integrated circuits SPACE switches are gates which are opened or closed according to a programme map contained in a memory. The programme may be fixed, i.e. wired in, or may be controlled by a computer.

The SDM is ideal for PCM switching, for it can transfer the data in its pulse time division form. Furthermore, space division enables the transfer of PCM data in parallel. The computer can process parallel information faster than serial information. Most modern computers perform all their logical operations by handling whole words in parallel. The increase in speed allows a large number of parallel words to be multiplexed into one common channel of the input to the computer. The serial to parallel conversion

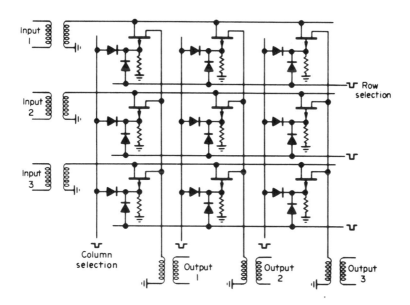

Fig. 3.15. Fully electronic (FET) SDM matrix.

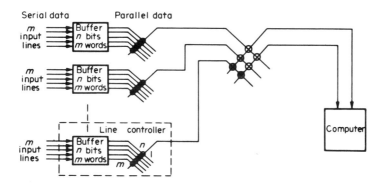

Fig. 3.16. Super-space division multiplex.

is performed in a line controller (as shown in Fig. 3.16). The lines transfer the serial information in a serial form and there it is buffered in a word-length register. The output from the buffers is multiplexed into a common parallel highway. This operation uses space-division multiplex as a parallel time-division multiplex. The circuit for this operation is the one that was shown in Fig. 3.6 for TDM, where the output of all the registers passes through a set of gates, except that the same circuit must be repeated for each bit so as to produce the parallel operation. The output from a number of line controllers is then passed through a second space-division multiplex to the computer. This second switch also operates as a parallel time-division multiplex, an operation which is called Superspace multiplexing. By using space-division multiplex for the computer input, three points are gained: the number of input lines is reduced considerably, the information presented to the computer is in parallel form, and the slow speed of the line

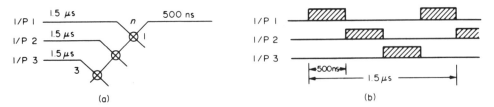

Fig. 3.17. SDM operates as TDM.

transmission is fed into the computer at a relatively high speed. It is important to remember that the input circuitry to the computer could be rather expensive and that this is reduced when superspace division multiplex is used.

The operation in increasing the speed of the data transfer can be seen from the example shown in Fig. 3.17. In this example the SDM operates as a TDM, that is, each cross-point is operated in turn. The time slot for each input line is enough for a complete word. In the case of superspace multiplexing the time slot need only be wide enough for a bit transfer, as the whole word is transferred in parallel.

3.8 CONCLUDING COMMENTS

The designer of telemetry schemes is confronted with a diversity of different modulation and multiplex programmes. Figure 3.18 shows the relationship of the different programmes and the alternative design paths available to the designer. The selection of the path to be used depends only on the system it is intended for. For Command and Control systems, the data transferred is usually in digital form, as marked with a thick line.

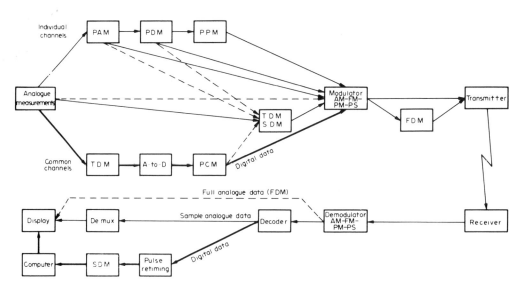

Fig. 3.18. Alternative design paths for telemetry schemes.

When transmitting the data on a communication path, it must be modulated (as will be explained in the following chapter). A number of modulation techniques are available, such as Amplitude Modulation, Frequency Modulation, Phase Modulation, or Phase Shift. The type of the transmission modulation techniques has no bearing on the original analogue waveform or the pulse modulation used. The modulation technique is a function of the efficiency of the transmission path and not of the telemetry scheme.

If frequency-division multiplexing is to be used, the basic frequency band of the measurement output must be modulated. Any modulation technique can be applied, although in all cases the side bands must be limited.

Time-division multiplexing can be applied only to pulses, and this demands that it is performed before the transmission modulation. There are systems where both TDM and FDM are used, one commoning the pulses orientated at a single location and the other commoning the transmission originating from a number of locations. The number of systems that this applies to is more than is realised, since most of the trunk telephone network uses FDM.

Space-division multiplex can be applied both to the analogue waveform and to the coded pulse train.

3.9 REFERENCES

1. Foster, L.E., *Telemetry Systems*, John Wiley & Son, 1965.
2. Hamsher, D.H. (Editor-in-Chief), *Communication Engineering Handbook*, Chapter 7 by P. Schneider, "Switching Engineering of switched systems", and Chapter 10 by M.E. Cookson and E.M. Thompson, 'Multiplexing', McGraw Hill Book Co., 1967.
3. Martin, J., *Telecommunications and the Computer*, Prentice-Hall Inc., 1969, pp. 265-292.
4. Fletcher, W., "Telemetry as a tool", *Electronic Equipment News* (Jan. 1968).
5. Notley, J.P.W., "Minicomputers for industrial automation", *Automation* (Sept. 1972), pp. 15-20.
6. Sheingold, D.H., "Analog-to-digital and digital-to-analog convertors", *Digital News* (Dec. 1972), pp. 50-56.
7. Chatelon, A., "PCM telephone exchange switches digital data like computer", *Electronics* (3 Oct. 1966).
8. Rose, D.J., "Pulse code modulation tandem exchange field trial model", *Electrical Communication*, Vol. 46, No. 4 (1971), pp. 246-252.
9. Pollard, J.R., "Technological progress in telecommunication switching", *Electronic and Power* (IEE) (Aug. 1970), pp. 305-308.
10. Digital Equipment Corporation Publication, *Logic Handbook, 1968*, Part V, "Analog-digital conversion", pp. 357-401.
11. *The Telemeter*, an Electro-Mechanical Research Inc. Publication;
    (a) No. 1, September 1965, "A Primer of Telemetry";
    (b) No. 2, December 1965, "An Introduction to Pulse Code Modulation";
    (c) No. 5, November 1966, "Recent Trends in Telemetry Data Processing".
12. *The Demodulator*, a Lenhurt Electric Publication; (a) February 1968, "Supervisory Control"; (b) March-April 1968, "Pulse Code Modulation"; (c) March 1971, "PCM Signaling and Timing".

CHAPTER 4

Data Transmission

4.1 INTRODUCTION

The successful operation of a Communication, Command and Control system depends essentially on its ability to transfer data between remote geographical locations speedily and correctly. This chapter discusses the basic techniques required for communication through the media of voice telephone lines.

Conveyance of data over lines has been in operation for over a century, providing a variety of services. It started with the invention of telegraphy and continued through the development of various teletypewriter systems. Although these systems are still widely used, they are much too slow for computer communication. The need for extremely fast communication systems has become of vital importance with the introduction of on-line real-time systems, where the terminals are spread over wide geographical areas. The increasing traffic problems calls for drastic adaptations following, and the introduction of new technologies. The technologies developed primarily for computer-type communication are known as Digital Data Transmission.

Most of the data transmission is conducted over the traditional telephone lines, apart from a few specially designed lines with wider bandwidth. The standard telephone lines are limited by their upper frequency, since they were designed to carry only voice traffic, that is, only analogue-type signals. In order to transmit digital data over these lines, the data generated by the computer or by the terminals must be converted into analogue signals so that they can be transmitted in their encoded form over the ordinary telephone lines.

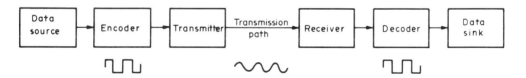

Fig. 4.1. Basic elements of digital data transmission.

Figure 4.1 illustrates the basic elements required for digital data transmission. The data to be transmitted must first be encoded into a form suitable for specific transmission handling, as the transmission over a communication line produces attenuation and phase delays; it is impractical to transmit the digital pulses in their raw form over telephone lines.

As already mentioned, it is necessary to modulate the data to be transmitted over analogue telephone lines and to demodulate the signal at the receiver end. Similarly as to

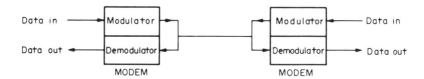

Fig. 4.2. The modem system configuration.

speech, data transmission is generally possible in both directions, although not simultaneously. The data transmitter–receiver which performs the dual process of MODulation and DEModulation is consequently referred to by the abbreviated form of MODEM (see Fig. 4.2). The modem performs the operation of translation between the binary data pulses and the voice frequency waveforms; hence it could also be regarded as analogue to digital and digital to analogue converters. The modem set also performs a number of control functions which are required to coordinate the flow of data between the terminals.

Fig. 4.3. Point-to-point simplex transmission.

Without regard to the communication media, there are three types of transmission.

(a) Simplex transmission, where a line carries data in one direction only, as shown in Fig. 4.3. This type of data communication is only applicable for data-collection systems and merely requires a modulator at one side and a demodulator at the other side.

Fig. 4.4. Point-to-point half-duplex transmission.

(b) Half duplex transmission, where a line can carry data in either direction but only in one direction at a time, as shown in Fig. 4.4. Here identical modems are inserted at both ends of the transmission path. This is a most popular data transmission configuration, as it can be used in any public telephone network.

Fig. 4.5. Point-to-point duplex transmission.

(c) Duplex (or full duplex) transmission, where a line can carry data in both directions simultaneously, as shown in Fig. 4.5. This double direction can be achieved by either transmission over two different frequency bandwidths or by a 4-wire line circuit.

Although data can flow in both directions simultaneously in a duplex configuration, it is common practice to operate it in a half duplex mode. In these applications the digital information flows in one direction while conversely the control data flows in the other direction to indicate either an acknowledgement or a request for retransmission of the message. This mode of operation (shown in Fig. 4.6(a)) saves time when two terminals or two computers communicate. It must be appreciated that for most communication procedures an answer-back is essential. (This will be discussed in greater detail when polling is introduced.)

A similar type of duplex transmission is operated in half duplex mode, as shown in Fig. 4.6(b). Here the data information is transmitted at a high speed on one of the channels, concurrently with the transmission of supervisory information in the opposite direction at the low speed of 75 bits/s. Although this system resembles a duplex transmission, it must be regarded as a half duplex system, since the actual information can flow in only one direction at any given time. In this system, the supervisory channel is adopted for alarm notification, when the receiver is out of synchronization with the transmitter, and for instruction to convert over from data to speech.

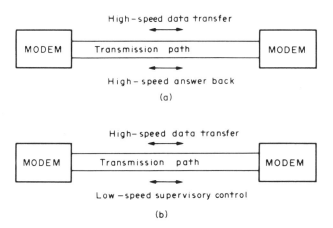

Fig. 4.6. Duplex transmission operating in half duplex mode.

4.2 PULSE CODE FORMATS

There are many different types of pulse waveform code formats which may be used in binary data transmission, each conveying digital information in a distinctive form as required for particular applications. All the code formats could be divided into three classes. The first criterion of division is the form of information transmission, viz.

(a) Full binary transmission, where both the '0' and '1' bits are part of the formats.

(b) Half binary transmission, where only the '1's are transmitted, having the '0's recognized by the absence of a pulse at the time of clock.

(c) Multiple binary transmission, where ternary and quadric codes are used for each transmitted pulse.

A second criterion of division is that of relation to the zero level, viz.

(a) Return-to-zero (RZ), where there is a return to the zero level after the transmission of each bit of information.

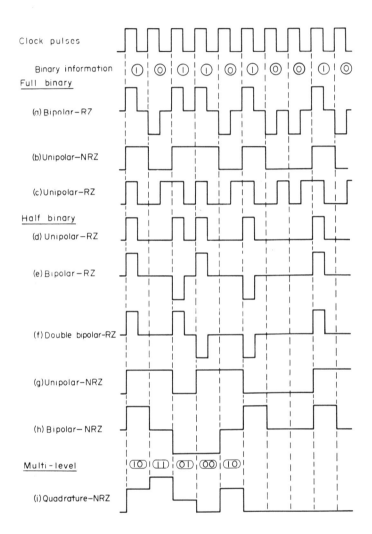

Fig. 4.7. Pulse code formats used in data transmission.

(b) Non-return-to-zero (NRZ), where there is no voltage level change if consecutive bits are transmitted, although there is a level change when there is an information variation from '0' to '1' or '1' to '0'.

A third accepted criterion of division is that of direction, viz.

(a) Unipolar, where the pulses are in a single direction.

(b) Bipolar, where the pulses are in both directions.

The various coding patterns for the respective pulse formats are illustrated in Fig. 4.7, where each pattern represents the same binary information of 1011010010. (These are explained below, where the prefix corresponds to that shown in Fig. 4.7.)

Full binary transmission

(a) Return-to-zero, bipolar, where opposite polarity pulses are used to transmit the '1' and '0' bits. Between the pulses no power is transmitted, i.e. there is a space between each pulse. This coding method is most popular for slow-speed transmission and is used mainly for FSK of speeds up to 600 bits/s. It is most reliable, since mutilations of the pulses during transmission are unlikely owing to the fact that the change in polarity is improbable without other obvious causes.

(b) Non-return-to-zero, unipolar, where the full pulses are spread out in time so that they occupy a full time slot, thus permitting acceleration of transmission. The term NRZ is applicable here because the pulses lose their individuality in the process of transmission and do not return to zero between successive pulses of '0's and '1's. This is a most popular data-transmission format, being used for most serial computer applications and for data transmission of speeds of 600/1200/2400 bits/s. The duration required for each bit of transmission is considerably reduced on average, thus resulting in higher transmission speeds. This code exploits the transmission bandwidth efficiently as the entire bit period contains signal information.

(c) Return-to-zero, unipolar, which has a symmetrical format and so does not contain a d.c. level. The code assumes a zero level crossing for each bit period, a feature which simplifies synchronization although at the expense of increasing the total bandwidth required. This code, which is also referred to either as a Manchester, split-phase or bi-phase code, is one in which a polarity reversal must occur during each bit period regardless of the binary information. The reversal occurs during the split period where the '1' reverses from positive to negative and the '0' from negative to positive. Owing to the reliable synchronization characteristics, this type is used for airborne to ground slow-speed data transmission.

Half binary transmission

In all the following formats only the binary '1's are represented by a pulse or a polarity change, while the binary '0's are seen as spaces. This is based on the statistical assumption that the number of '1's in a pulse train is equal to the '0's, resulting in a reduction of the transmission power with the possible increase of the transmission speed.

(d) Return-to-zero, unipolar, which is the same as the code shown in (a) except that the '0' pulses have been eliminated. This code format modifies the frequency spectrum by reducing the high-frequency signals which cause cross talk. However, it also produces a d.c. component which is difficult to transmit. This coding method wastes 50% of its bandwidth, since only half its bit period contains bit information.

(e) Return-to-zero, bipolar, which is the same as the code shown in (d) except that the polarity of each consecutive pulse is reversed. By inverting the alternate '1' pulses the d.c. component is removed and the inter-pulse interference is reduced. This code also facilitates error detection, since all that is required is to remember the polarity of the last pulse received. However, with the '0's not transmitted it gives no indication of the location of the error. This code format is most popular for PCM transmission at very high speeds. An alternative name for it is pseudoternary.

(f) Return-to-zero, double bipolar. This code, suggested by M. Karnaugh, extends the RZ bipolar code shown in (e) with the difference that the polarities are inverted only after every two consecutive '1' bits. The format further reduces the inter-symbol interference, thus promoting better utilization of the bandwidth.

(g) Non-return-to-zero, unipolar, where the polarities are reversed each time a '1' bit is transmitted. No polarity reversal takes place when a zero is transmitted, regardless of the direction prevailing as the last '1' bit. This method, known as NRZI code format, is widely used for applications of tape recording but it may also be used for data transmission purposes.

(h) Non-return-to-zero, bipolar, which is a three-level code format known as Duobinary code. Although it looks like a RZ code it does not accord with the definition, since the pulses do not return to zero after each data pulse. Here the binary '0' is represented by the zero level and the binary '1's by a bipolar form. The polarity of the binary '1's is reversed only when the number of the consecutive binary '0's between the '1's is odd. If the consecutive binary '0's total up to an even number, the two consecutive spaced '1's remain at the same polarity. This special code is intended for three state transmissions, as will be explained in Section 4.6.

Multi-level transmission

The allowable bandwidth controls the number of symbols per second which can be transmitted. In most data transmission systems hitherto discussed each bit transmitted corresponds to one symbol. By having one symbol signal to represent a number of bits, the speed of transmission could be increased. However, a penalty must be paid for the higher data rate, as the data is more vulnerable to noise and other distortions. This is because individual states are more difficult to distinguish in a multiple level method than in a two level method.

There are many multi-level code formats; for example, a ternary code which replaces the binary code. This code could be transmitted by a three level as in a NRZ code. This method is not widely used, because the logical circuitry involved is rather complicated. A more popular coding method is the quadrature binary code format.

(i) Non-return to zero, quadrature binary, is a four-level code where each level represents two binary bits, that is, a bit pair 00, 01, 10, 11 respectively. The 2-bit pair is referred to as a 'dibit'. As illustrated by the pulse train in Fig. 4.7 the pulse sequence duration for a dibit can be transmitted at the time of one bit pulse. In practice, however, the dibit is transmitted at a speed less than double that of a bit, since the speed is limited by the available bandwidth and the received threshold.

Similarly to the quadrature binary, eight and sixteen levels could be used to achieve higher speeds. In fact, a speed of 9600 bits per second is achieved with an eight-level code, but it must be taken into account that this method is susceptible to line noises. In view of this reservation the maximum speed is reached only on specially selected lines which use modems with automatic equalizers.

Before concluding this section it is important to clarify and define the terms used for the measurement of the rate the data is transmitted. If wrongly applied, the terms can be grossly misleading.

A "BAUD" is a unit of signalling speed equal to a code element per second. It refers to the number of times the line condition changes per second. The individual code elements could be of equal or varying length, as shown in Fig. 4.8(a). This code, which derives its name from its inventor Baudot, is used mainly for telegraph transmission.

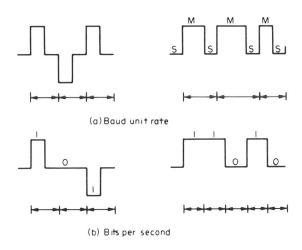

(a) Baud unit rate

(b) Bits per second

Fig. 4.8. Unit of measuring speed of transmission.

It is obvious from the definition that the use of 'baud' for data transmission could be confusing, as it does not represent the actual information which is transmitted. For all the code formats discussed here, except that of full binary, bipolar, RZ, the wrong interpretation could have serious consequences.

For data transmission, the unit of signalling speed should be measured by the number of bits transmitted per second. The term 'bits per second' (bits/s) refers to the actual number of binary digits that are transmitted per second. This code does not refer to the number of pulses transmitted but to the number of information units transmitted, as shown in Fig. 4.8(b). In NRZ-Unipolar, 2 bits per second are equal to 1 baud. In a dibit code format each pulse refers to 2 bits.

4.3 TRANSMISSION MODULATION

The transformation of the digital data into useful analogue signals for transmission is accomplished, as previously stated, by modulation techniques. There are three types of modulation techniques which are used for data transmission: Amplitude modulation, Frequency modulation and Phase modulation. A fourth modulation technique could be added, formed by the combination of amplitude modulation with that of phase modulation. In each of these modulation techniques a sine wave carrier is employed to convey the data by means of a change of its amplitude, its frequency or its phase. In principle, the three modulation techniques follow the block diagram shown in Fig. 4.9.

The technical aspect of modulation techniques has been adequately covered in the literature. The treatment discussed in this book deals with the system design aspect required for communication networks rather than with the circuit design of the modems.

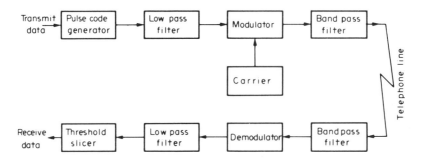

Fig. 4.9. Block diagram of a data-transmission system.

In designing data transmission networks the main aspect is the efficient utilization of the transmission path. In effect this means transferring the maximum amount of data with negligible errors, at a reasonable price, and using the full bandwidth available.

Nyquist in 1928 showed that the minimum bandwidth necessary to transmit N binary elements per second is $N/2$ cycles of the bandwidth. In other words, the numerical speed rate is equal to twice the line bandwidth. For data transmission over the voice telephone lines, the available frequency bandwidth is from 300 c/s to 3300 c/s. Thus, theoretically, the maximum speed that can be transmitted on a telephone line is 6000 signal elements per second. In practice, however, the results are discouragingly different. Modern transmission techniques require about twice the Nyquist bandwidth to achieve the above speed. Nevertheless, faster speeds could be accomplished by suitably coding the data, using one signal element to signify a number of bits.

The aim in designing is to approximate the Nyquist transmission value. This can be attained by minimizing the influence of noise and controlling the intersymbol interference. The noise itself is an external property, the effect of which could be curtailed by increasing the available received threshold obtained after the demodulator. The intersymbol interference depends on the shape of the pulse which is fed into the modulator. The ideal rectangular sharp cut-off pulse is not really sought, as the tails of the pulse cause high peaks. A relatively slow rise and fall time pulse is required for best results, which can be achieved by low pass filters placed between the data input and the modulator, as shown in the block diagram of Fig. 4.9.

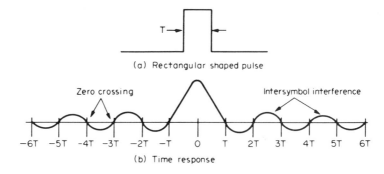

Fig. 4.10. The impulse response of a rectangular pulse.

A rectangular shaped pulse, as shown in Fig. 4.10 (a), fed into a low-pass filter, has the ideal time response of sine x/x, as shown in Fig. 4.10 (b). Nyquist demonstrated that the transfer characteristics have a symmetrical impulse response on both sides of the pulse which gradually "roll off" to zero. These impulses are called intersymbol interference. Each impulse passes the zero level at intervals of T, where T refers to the pulse width and $1/T$ is the number of bits per second that can be transmitted. The low-pass cut-off frequency according to the Nyquist theorem is $f = 1/2T$, which in theory enables the transmission of $2f$ bits per second without interference between the peaks of adjacent pulses.

As shown, the spectrum of the filtered pulse gradually rolls off to zero with polar symmetricy at T intervals. Nyquist demonstrated that the amplitude of the roll-off signals could be reduced by modifying the shape characteristics of the pulse without disturbing the zero crossing-points. This effect could be achieved by means of a low-pass filter having a roll off cosine characteristic (as shown in Fig. 4.11).

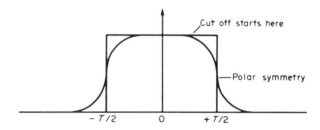

Fig. 4.11. Modified pulse shape using cosine rolloff.

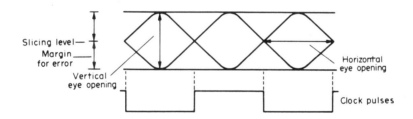

Fig. 4.12. Eye pattern for a two-level code.

A useful concept in evaluating the intersymbol interference and hence the quality of the received demodulated data, is the "eye pattern". The eye pattern shown in Fig. 4.12 is formed by superimposing a random pulse train. The intersymbol interference is evaluated by the available threshold, which is defined as the interval at the centre of the eye opening. When the received signal is disturbed beyond the specific threshold, the state of the data is inverted and an error is recorded.

When transmitting a multi-level signal, such as a quadrate amplitude modulations signal, the eye pattern consists of three levels, as shown in Fig. 4.13. The vertical eye opening remains relatively unaffected but the horizontal eye opening deteriorates rapidly with the increasing number of levels. This reduction in the horizontal threshold is due to the many transitions between the levels.

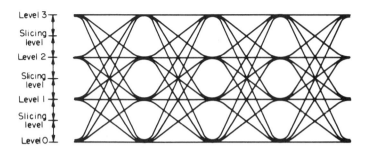

Fig. 4.13. Eye pattern for a four-level code.

4.4 AMPLITUDE MODULATION (AM)

In amplitude modulation the carrier is varied in accordance with the amount of information transmitted. The resultant wave varies in a direct relation to the modulation wave, producing two identical symmetrical sidebands around the carrier. As the same information is contained in the two side bands, only one side band is required to transmit a message. Sideband suppression is obtained by suitable filtering. A single side-band-suppressed carrier is the most widely used AM technique, as will be explained later.

For data transmission represented by 1 and 0 bits, the carrier is simply switched on or off respectively (as shown in Fig. 4.14). The amplitude modulation is also referred to as ON/OFF keying. This is a 100% modulation which represents the optimum signal-to-noise requirements.

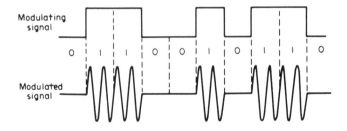

Fig. 4.14. Amplitude modulation of a rectangular pulse.

Since amplitude modulation is extremely vulnerable to common type noise, its application is rather limited. In this form of modulation, attenuation, delay distortion and echoes will change the shape of the signal at the receiver output. In addition, cross talk from adjacent lines might be superimposed on the received signal.

The speed of transmission may be increased by transmitting four levels of amplitude instead of the two levels of ON/OFF information, as shown in Fig. 4.15. In this case each of the four levels represents two digits simultaneously, forming what is called a dibit. Theoretically the speed of transmission should be doubled, but in practice the gain in speed reduces the threshold available for detecting the pulse state. In consequence, this type of four-level modulation is more susceptible to transmission errors, making its application somewhat limited. Nevertheless, the principle of this multilevel modulation

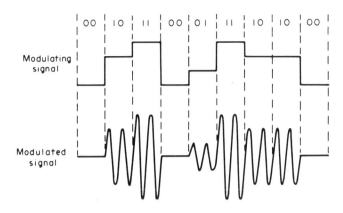

Fig. 4.15. Amplitude modulation of a four-level pulse.

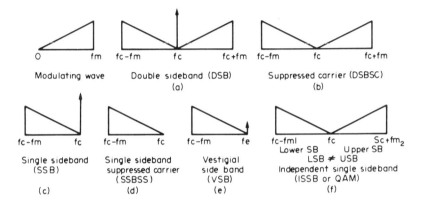

Fig. 4.16. Various amplitude modulation transmission techniques.

has been applied to other types of modulations (as will be discussed later on).

In view of the fact that each side band contains the same information, amplitude modulation could be transmitted in six forms. These AM transmissions media are shown in Fig. 4.16.

(a) Double sideband (DSB), which was used in the very early data transmission low-speed equipment but has since been replaced by FM methods. In this type of data transmission the '1' bit was represented by a continued amplitude wave of fixed frequency. The bandwidth required for transmission of the information with a basic band of f_m frequency is $2f_m$. This technique has the disadvantage that much of the power is lost in the transmission of the carrier and it conveys no useful information. For ON/OFF amplitude modulation (i.e. 100% modulation index) the power in each sideband is only 25% that of the carrier.

(b) Suppressed carrier (DSBSC), which is a form of double sideband with the carrier removed. The bandwidth required for transmission still remains $2f_m$, but it is possible to increase the sideband power and still keep the transmitted power below that of a double sideband.

(c) Single sideband (SSB), in which one of the sidebands is removed by suitably filtering one of them. This suppression of one sideband can be performed without destroying any of the information intelligence. Either sideband could be used, depending on the application. This technique makes better use of the available frequency spectrum but is less tolerable to delay distortions, amplitude variations and frequency or phase shifts. The power transmitted is divided between the carrier and the single sideband. This is an important factor in data transmission, since a large amount of power is not useful for transmitting the actual information.

(d) Single sideband suppressed carrier (SSBSC), where all the available transmission power is concentrated in the signal sideband which contains the desired information. The carrier can be suppressed by the application of a balanced modulator and removing one of the sidebands with the aid of a filter. This technique, in fact, preserves the original bandwidth (though shifted in frequency). As explained in the previous section, the spectrum of a pulse signal gradually rolls off to very low frequency, including d.c. This d.c. and low frequency cannot be suppressed with the filters. Although this system has better frequency utilization it is not economically practical for data transmission, as the circuitry required is very complex. It is used for high-speed data transmission (PCM) where the advantages outweigh the disadvantages.

(e) Vestigial sideband (VSB), which is a variation of SSB but also combines some DSB properties. This technique is popular for high speed modems for the transmission of two signals simultaneously. The reason for the attraction is that it permits economy both in the transmission power and bandwidth. In vestigial sidebands, both the wanted sideband and a portion of the unwanted one are transmitted. There is a partial suppression of one sideband in the vicinity of the carrier and a partial suppression of the other. This leaves one sideband with only a vestige of the carrier which serves frequency accuracy and stability. The carrier is reduced to less than half its full DSB value and this releases more power to the sideband transmission. The VSB is used for the transmission of four-level data by employing two modulators each having a carrier with a $90°$ phase between them. VSM techniques require a bandwidth approximately 1.3 times that of an SSM technique.

(f) Independent single sideband (ISSB), which is based on two independent single sideband amplitude modulators with suppressed carrier. In one modulator the lower sideband is eliminated, while in the other the upper sideband is eliminated. Each modulator is switched ON or OFF according to the data, where two bits are fed in parallel, one to each modulator. The two SSBs could then be combined and transmitted in quadrature. This technique is known also as quadrature amplitude modulation (QAM). Here two double sidebands operate in a quadrature phase which can be separately and coherently demodulated. This is possible because the double sideband signals have no quadrature component and so there is no interference between the information transmitted in both combined sidebands.

Multi-level data transmission is the goal of most of the modern development today. It can be achieved by three amplitude modulation techniques. The simplest one uses a four-level amplitude, but this is not widely adopted owing to its susceptibility to noise. The other two are QAM and VSB, which enable the speeds to reach 9600 bits/s.

4.5 FREQUENCY MODULATION (FM)

In frequency modulation the carrier frequency is changed according to the

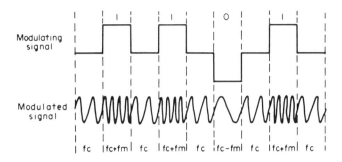

Fig. 4.17. Frequency modulation of a RZ pulse sequence.

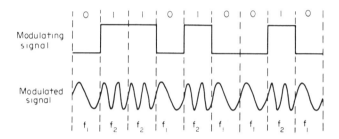

Fig. 4.18. Frequency modulation of a NRZ pulse sequence.

information transmitted. The ON/OFF data information shifts the transmitted frequency in one direction to indicate a '1' and in the other direction to indicate a '0'. The amount of frequency shift in both directions is the same. Digital frequency modulation is also known as frequency shift keying (FSK).

Frequency modulation is less vulnerable to amplitude level changes or to impulse noise, and hence is superior to amplitude modulation, but not with respect to the data rate achieved over a given bandwidth. It was first introduced to replace amplitude modulation, although it generally requires a larger bandwidth. It is used today in modems operating at speeds up to 1200-1800 bits/s and requires approximately the same bandwidth as a double-sideband amplitude modulation.

Frequency modulation is a continuous analogue process with no changes in its amplitude transmitted level, where in the absence of information the carrier is transmitted. This is one of the reasons why FM is less insensitive than the basic digital AM systems to noise, cross talk and intermodulation. A further advantage for the continuous transmission of the carrier is that the received signal-to-noise ratio is constant.

Two basic types of code formats can be used for frequency modulation, one based on the RZ code with three frequencies and the other on NRZ code using the two side frequencies. In return-to-zero, shown in Fig. 4.17, the two discreet frequencies are used to transmit the binary states, while the carrier Fc is used to transmit the intervals between the pulses. Fc−Fm represents the '0' binary state and Fc + Fm represents the '1' binary state. In non-return to zero the frequency is shifted between two frequency values (as shown in Fig. 4.18). The higher frequency is assumed to have the value of '1' binary state and the lower frequency as the '0' binary state. With no information transmitted, the upper frequency is usually transmitted continuously. When varying the frequency, between 1 and 0 states, the phase must be preserved. This can be achieved by using a single oscillator and adding or deleting a capacitor to change the frequency of oscillation.

The three-frequency FM technique is used for telemetry applications and for some manually operated terminals. This type of modulation has been mainly adopted for speeds up to 150 bits/s due to the loss of time period between adjacent pulses. The two-frequency FM technique is much more widely used, despite the bandwidth restrictions. It is chosen for its simplicity and economy, factors which are more important than bandwidth efficiency.

For many control applications the three-level technique is very popular, because the third available state could be used to transmit additional information. The three-level transmission also simplifies the synchronization of the receiver with the transmitter and the reconstruction of the clock pulses. The latter is possible because in a RZ code a pulse is transmitted for each clock pulse. Although the three frequency transmission is primarily used for slow speeds it could also be applied for ternary code transmission, possibly increasing the speed up to 2400 bits/s. (This type of transmission uses a duobinary code and is discussed in the following section.)

4.6 DUOBINARY TECHNIQUE

The duobinary technique code format transforms the conventional binary data into a three-level waveform in a manner which allows doubling the transmission rate. While this coding technique could be applicable to AM, FM, or PM, it is generally used only with FM, since it already possesses the three levels and their utilization does not increase the bandwidth. The 'duo' in the name refers to the doubling of the speed and not to the levels involved.

The logical principle of the duobinary transmission is illustrated in Fig. 4.19. The binary data input, in the form of a NRZ train of pulses, is transformed into a specially coded bipolar RZ pulse train. The '0' data bit remains at the centre level, while the polarities of the '1's are reversed, provided that there is a '0' bit between them and that the number of the consecutive '0' bits is odd. The polarity of the '1's remains in the same direction if there is no '0' bit between consecutive '1's or if the number of the consecutive '0's between the '1's is even.

Figure 4.19 also illustrates the analogue signal line obtained from the binary input

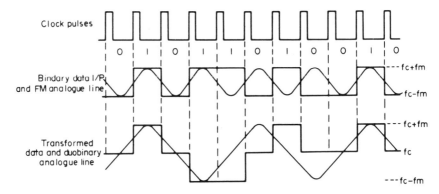

Fig. 4.19. Duobinary coding principle.

pulses and the analogue signal line obtained from the transformed duobinary pulses. By comparing the time responses of the two analogue signals it can be seen that the basic frequency rate of the original signal is twice that of the coded signal. This accounts for the fact that duobinary transmission can be transmitted at twice the speed of the standard transmission.

When using duobinary transmission with three-level frequency modulation, the carrier frequency is allotted the '0' binary state which represents the zero centre level of the transformed duobinary pulse train. Frequencies Fc + Fm and Fc–Fm are then allotted to represent the positive and negative levels respectively. Returning to the pulse waveforms in Fig. 4.19, it can be seen that in the original waveform there could be a change from frequency Fc + Fm to frequency Fc–Fm within a time period allocated to one bit. In duobinary transmission this transition must take at least two time periods, as each transition must pass a time period with the transmission of the carrier frequency Fc. The result of transformation from the unipolar to the biopolar pulses is the redistribution of the spectral density of the original data into a highly concentrated energy density at low frequencies.

The generation of the three levels is quite simple, as shown in Fig. 4.20. The inverted

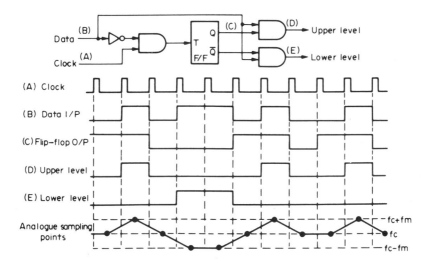

Fig. 4.20. The duobinary conversion process.

data is fed into a T-type flip-flop which changes its state for each '0' input pulse. (If the carrier frequency is transmitted as the '1' state then the flip-flop must be inverted for each '1' input pulse.) The data must first pass through a clocked gate, to distinguish between consecutive '0' pulses. The two complemented output pulses obtained from the flip-flop are then added separately to the original data, thus providing the upper and lower levels. The centre level is obtained either from the original data or from the zero level of the combined upper and lower levels. Also shown in Fig. 4.20 are the analogue sampling points obtained from the three levels.

The transformation back to the original NRZ pulse sequence is obtained by simply reversing all the negative going pulses to positive going pulses. At the receiver this is

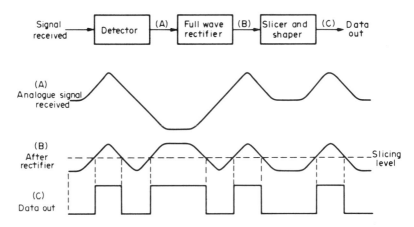

Fig. 4.21. Duobinary receiver process.

achieved by passing the waveform obtained from the demodulator through a full wave rectifier, then slicing and shaping the pulses (as shown in Fig. 4.21).

4.7 PHASE MODULATION (PM)

In phase modulation the phase of the carrier frequence is changed according to the data transmitted (as shown in Fig. 4.22). This modulation is also known as phase shift keying (PSK). In phase modulation both the frequency and the amplitude transmitted remain constant throughout the transmission period. In this respect PM is superior both to AM and FM techniques and so replaces them for speeds above 1800 bits/s. The main disadvantage of PM is the complexity of the equipment required, although the cost is justified wherever high data rates are required. The high speed is achieved by transmitting multi-states for each phase shift. In the illustration shown in Fig. 4.22 the carrier was shifted by 180° for each change from the '1' to '0' or '0' to '1' binary states. Nevertheless, phase changes of 90° and 45° are feasible and practicable and so are used for multi-level transmission.

Phase modulation is primarily used for digital transmission and is not intended for voice transmission or for telemetry analogue information. This is because the phase changes cannot be detected with accuracy. Although it is difficult to detect small phase changes required for analogue transmission, it is still possible to transmit small phase changes. Phase changes of 90° or even 45° could be transmitted, and be easily detected.

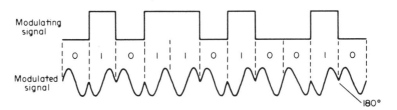

Fig. 4.22. Basic principle of phase modulation.

With the 90° phase variation, four-phase changes are obtained, allowing two bits to be represented by each phase change. In this case two bits of information can be transmitted simultaneously, as in quadrate multi-level transmission. With 45° phase variations, eight-phase changes are obtained, allowing three bits to be transmitted simultaneously.

When a four-phase change system is used, two modulators are operated in parallel, each shifting the phase by 180°. The output of the two modulators is in phase quadrature, combining in the output circuit and giving the resultant vectors shown in Fig. 4.23. There are two choices for the use of phase shift applications. The four-phase shifts could be ±180° and ±90°, as used outside the United States. Alternatively, the phase shift choice could use ±45° and ±135°. The latter is used in most four-phase modems in the United States and presents performances superior to the first choice. Table 4.1 shows the codes employed for each transmitted dibit information using both alternative phase shift choices. The angle stated is the phase shift occurring in reference to the steady state of the previous dibit.

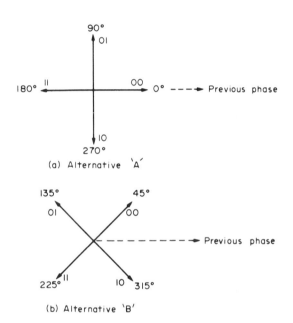

Fig. 4.23. Vector diagram of choices of quadrate phase shift.

TABLE 4.1

Transmitted dibit	Phase shift in reference to previous dibit	
	Alternative A	Alternative B
00	0°	+ 45°
01	+ 90°	+135°
11	+180°	+225°
10	+270°	+315°

From Table 4.1 it can be seen that when transmitting a continuous pulse train of '0's there is no phase shift when alternative A is used. In this arrangement synchronization could easily be lost when a long string of continuous binary '0's is transmitted. This could be overcome by a special scrambler which changes the data, thus obviating a long string of '0's. Alternative B has the advantage that each dibit must cause a phase shift. For a continuous transmission of '0's there is a repeated phase shift of 45° after each consecutive 00 dibit, and this arrangement is actively maintained.

An example of a four phase-shift transmitter is shown in Fig. 4.24. The example is

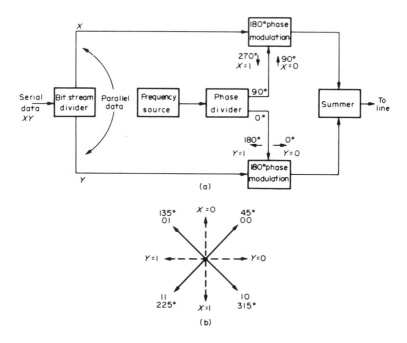

Fig. 4.24. Block diagram of a four-phase modulator transmitter.

based on the phase angles of alternative B but it could easily be applicable to alternative A. The data stream is fed into the transmitter in serial form and there it is converted into two parallel streams. Two consecutive adjacent bit pulses are always modulated simultaneously. Each modulator changes the frequency phase by 180° according to the specific data applied to it. By combining the two modulated waves a quadrate phase change can be generated out to the line. In the combining circuit the resultant vector presents the phase angle shift for each dibit, as shown by the vector diagram.

The two modulators are fed by the same frequency source except that they are perpendicular to each other. For alternative A the phase to the modulators is ±45°, while for alternative B it is 0° and 90°. The phase angle shift for each modulator and the summed vector phase angle transmitted into the line is given in Table 4.2.

The receiver has no absolute sense of phase and the decoding can only be obtained by reference to another angle. There are two basic techniques in detecting the relative phase shift.

The first method, termed fixed reference phase detection, is shown in Fig. 4.25. The

TABLE 4.2

Input dibit	Alternative A			Alternative B		
	1st mod.	2nd mod.	summed	1st mod.	2nd mod.	summed
00	45°	315°	0°	90°	0°	45°
01	45°	135°	90°	90°	180°	135°
11	225°	135°	180°	270°	180°	225°
10	225°	315°	270°	270°	0°	315°

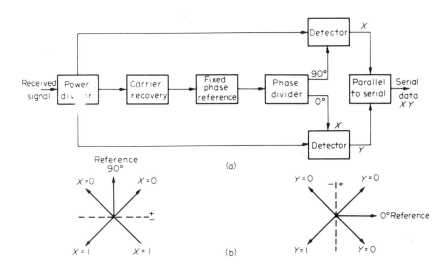

Fig. 4.25. Fixed reference detection of a four-phase signal.

fixed reference is based on the original carrier frequency which can be recovered from the transmitted waveforms, irrespective of the phase shift. This is accomplished with a full wave rectifier, which thus recovers the second harmonic (as will be discussed in greater detail in Section 4.9). As in the transmitter, the recovered carrier frequency is fed to two parallel detectors. The carrier is phase split into two waveforms which have a 90° angle between them. If the incoming phases are 0°, 90°, 180°, 270°, then the fixed reference is shifted by ±45°. If the incoming phases are 45°, 135°, 225°, 315°, then the reference is based on the relation to 0° and 90°. The received waveform is compared at the detector with the appropriate reference. If a positive indication is obtained from the corresponding detector, then a '0' is recorded for the respective bit of the reconstructed dibit. If a negative is obtained, then a '1' is recorded. Table 4.3 shows the relation between the input received phases, the polarity of the detector outputs and the dibit recovered data.

The fixed reference detection is susceptible to phase jitter and other channel disturbances. This creates the problem of both establishing the fixed reference and maintaining it in its fixed phase. Once the receiver loses synchronization it takes time before the synchronization can be established again. For high speeds the fixed reference is

TABLE 4.3

Input phase	Detected output +45°	−45°	Recorded dibit	Input phase	Detected output +90°	0°	Recorded dibit
0°	+	+	00	45°	+	+	00
90°	+	−	01	135°	+	−	01
130°	−	−	11	225°	−	−	11
270°	−	+	10	315°	−	+	10
	(a) Alternative A				(b) Alternative B		

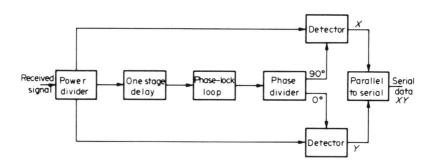

Fig. 4.26. Differential phase detection of a four-phase signal.

required to be accurate and stable so that it will be possible to distinguish between adjacent dibits. These problems could be avoided by using differential phase detection, but at the risk of incurring a penalty in the signal-to-noise ratio.

The second detecting method is based rather on the change relative to the last existing phase than to a fixed reference standard. This method is known as differential phase detection and is termed differential phase modulation (DPM or DPSK). In this modulation method each dibit is transmitted in reference to the previous dibit. (A simplified version of a DPSK receiver is shown in Fig. 4.26.) The principle of operation is similar to that of the fixed reference method except for the reference reconstruction. In this method each dibit is delayed and buffered for one dibit period and then compared with the following dibit signal.

4.8 MULTI-LEVEL TRANSMISSION

The speed of transmission can be increased by using three or four levels for each state transmitted, thus providing an 8- or 16-bit combination code. A wide choice of techniques is available for transmission speeds of 4800-7200-9600 bits/s. These techniques are based on vestigial sideband amplitude modulation, quadrature amplitude modulation and phase modulation. The combination of phase and amplitude modulation is a popular type of technique for achieving the higher speeds. Figure 4.27 illustrates means of achieving eight transmission states where each state represents three bits:

(a) shows an eight-phase shift modulation operation with a phase angle of 45° between each vector;

(b) shows a combination of phase and amplitude modulation where only four-phase shifts are used but each phase has two amplitude levels, thus giving a sum of eight states;

(c) this shows a combination of AM and PM, which is more efficient than the first two methods. Here eight phases are used, but with two different levels for each adjacent phase vector. This operation method presents the best tolerance to interference.

As explained at the beginning of the chapter, increasing the number of the levels reduces the horizontal eye openings and consequently causes an increase in the intersymbol interference. The transmission must thus suffer a noise penalty following the increase of the number of levels. The intersymbol interference could be minimized by shaping the pulses in the manner already described. For high-speed transmission, however,

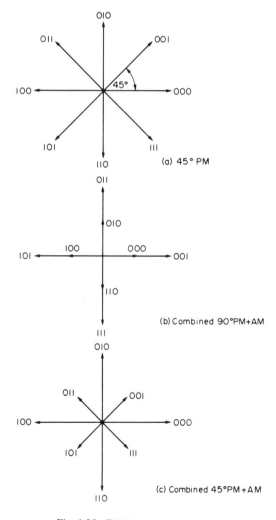

Fig. 4.27. Eight-level transmission.

more precise control of the intersymbol is required. This is achieved only by using automatically adjustable equalizers of the transversal filter type which automatically corrects both amplitude and delay distortions.

4.9 SYNCHRONIZATION

One of the main problems in data transmission is the ability to accurately reconstruct the transmitted data back to its original form. This can only be accomplished if the receiver operates in full coordination with the transmitter, namely, with the receiver in synchronization with the transmitter. The synchronization can be achieved in either of two general techniques, viz. asynchronous transmission, where each character is transmitted separately, and synchronous transmission, where a whole message is transmitted at a fixed speed. In both cases the data is transmitted in series over a transmission path.

Asynchronous transmission is favoured for low transmission speeds and synchronous transmission for high speed. In the latter technique the essence of the clock must be transmitted with the data in order to mark the location of the data bits received.

In data transmission there are four situations where the receiver must be interfaced with the transmitter: (a) bit, (b) character or frame, (c) fixed sized block, (d) message. The first two apply to both synchronous and asynchronous transmissions, while the last two apply only to synchronous transmission. Different techniques are used for interfacing each aspect. The bit interfacing is usually performed either by measuring the pulse duration or by reconstructing the clock pulses. The character or frame is usually reconstructed by first recognizing the start of the character pulse sequence, achieved by the transmission of a special start code. The end of the character is recognized either by a special stop code or by counting the bit stream. The block and message interphasing is accomplished by transmitting a special code, depending on the format used, stating the start of block or message (SOB or SOM) and another code specifying the end of the block or message (EOB or EOM).

Both synchronous and asynchronous techniques require special equipment to perform the coordination between the sending and the receiving modems.

4.10 ASYNCHRONOUS TRANSMISSION

The asynchronous transmission usually operates on a character basis where synchronization must be re-established each time. It originated for telegraph and teleprinter type transmission, where each character is typed and transmitted separately, one after the other. Special start and stop bits represent the beginning and the end of each character; hence this technique is often referred to as Start–Stop transmission. Asynchronous equipment being usually manual, the maximum speed is specified and the equipment can operate randomly at any speed up to this limit.

Asynchronous equipment is generally associated with electromechanical equipment where the electrical currents are broken or maintained according to the data. It is customary to refer to the logical levels of continuous current as 'mark' (logical bit '1') and to a break in the current as a 'space' (logical bit '0'). Each character generally consists of 5, 7 or 8 bits, and is initiated by a start signal of a single space signal. At the end of a character there is generally a single parity bit followed by a mark signal, usually of 2-bit periods (as seen in Fig. 4.28). The stop mark signal is maintained until the next character is transmitted, hence, there is no upper limit to the stop pulse length, although there is a minimum to its lower limit, depending on the system used.

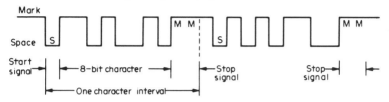

Fig. 4.28. Asynchronous data sequence.

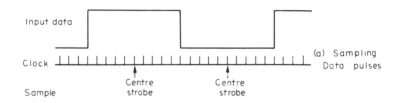

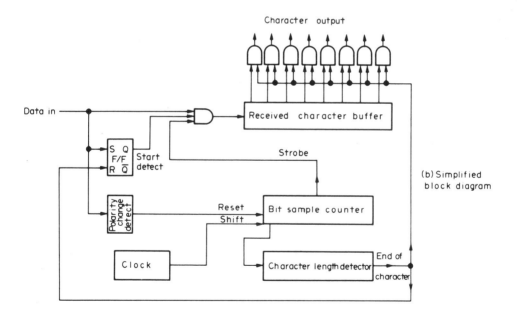

Fig. 4.29. Schematic asynchronous receiver.

The nominal bit interval is controlled by the maximum transmission speed, which is usually defined in bauds. No clock pulse is transmitted, but the individual data bits could be reconstructed at the receiver end by calculating the location of the bit centre. This is accomplished by a counter which possesses the characteristic of a full cycle count equal to the nominal bit pulse duration. The accumulation of all the data bit pulse displacements in the character must be less than the pulse duration. Thus the centre strobe pulse will always sample within a certain percentage of the centre of each bit. The transmitted pulses are relatively of equal periods but are distorted by noise and delays with consequent jitter in pulse duration.

A simplified schematic diagram of an asynchronous receiver equipment is shown in Fig. 4.29. The leading bit of the character is well defined and is used to initiate the full operation. The bit sample counter is reset for each consecutive bit, although if a continuous mark or space is transmitted the full counter cycle is in effect. Once the beginning of a character is detected, then the end of the character is easily determined by using a second counter which counts the number of bits in each character. After the last data bit is transmitted, the line remains at the mark signal level and all the elements are reset in preparation for the next character. The bits received are assembled in a buffer till the full character has been received. Before the full character is transferred to other locations, the data is usually checked for errors (this will be discussed again in Chapter 12).

Asynchronous equipment is simple and inexpensive, both to instal and to maintain. It could be used over a wide range of data rates and organized for diverse applications. However, it must be noted that asynchronous transmission is slow, limited in accuracy and inefficient. It requires separate timing circuits both at the transmitting and receiving modems, with about 25% of the data stream being redundant for the start and stop signals, apart from the extra bits that are required for synchronization and parity. Asynchronous transmission equipment is available up to 1200 bits/s, although most applications restrict the use of the technique for speeds below 600 bits/s.

4.11 SYNCHRONOUS TRANSMISSION

Basically, synchronous transmission is defined as a mode transmitted over a line at a constant speed. The term 'synchronous' transmission may be somewhat misleading, as it refers not only to continuous transmission of data but also to random transmission of messages. In asynchronous transmission each character is transmitted separately, while in synchronous transmission the units are large blocks of data. Another important aspect of synchronous transmission is that the clock information must be sent over from the transmitter to the receiver so as to coordinate the timings of both ends. Although there are a number of synchronous systems which possess different operational details, all conform to the above definition during the transmission period.

As already stated, synchronous transmission is used for higher speeds than those of asynchronous transmission. It also makes better utilization of the transmission path and requires less redundant data, thus resulting in a most efficient communication system. Lastly, synchronous operation implies that accurate timing sources are available both at the transmitting and receiving ends, with consequent better protection against errors.

The simplest synchronization method is to transmit all the timing synchronous data on a separate transmission path (as shown in Fig. 4.30). This could be accomplished either by a separate transmission line or by a separate frequency channel in an FDM system. This method can prove to be effective only in a wide-band system where a number of channels operate in parallel at high speed, with one of them reserved to transmit exclusively synchronous information.

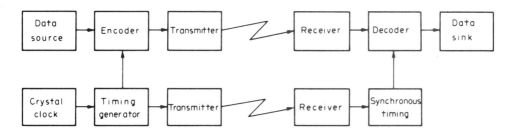

Fig. 4.30. Transmission of synchronous data over a separate channel.

In general, the process of synchronous operation is based on the transmission of fixed format data, composed of blocks which contain a few hundred characters in an NRZ unipolar stream. Each character typically consists of 7-8 bits, where 7 bits are associated with the transmitted information, while the last bit is available for parity or synchronization.

A block of data is termed in this book "a fixed size message". This is to distinguish the block from an unqualified message which could be variable in size. It must be noted that a block can consist of a number of messages or, conversely, a number of blocks may be required for the transmission of a single message. This distinction is important when considering a computer controlling the transmission of messages.

At the receiver the synchronous circuitry must be able to reconstruct the data transmitted. This is accomplished by extracting from the received signal the following elements: (a) clock timing, (b) data bits, (c) character start, (d) block start and end, (e) message start and end. Since the format is fixed and the transmission speed is relatively constant, the receiver can postulate logically all the required synchronization information. To achieve this, the format of the block and character must include some redundancy conveying sufficient information to assist in reconstructing the original transmitted data. It is essential for the receiver to know when to look for the synchronous bits and for the specially coded synchronous information characters.

The efficiency of a transmission system is measured by two factors:

(a) acquisition time, which is the time of the process for establishing synchronization;

(b) tracking time, which is the time of the process for correcting the synchronization drift.

The receiver must be able to extract enough data out of the received signal to reproduce the original transmitter timing, in order to correctly sample each bit received. In most applications the receiver consists of a local free-running oscillator which is pulled into synchronization with the clock by the aid of the received information (as shown in Fig. 4.31). There are several techniques for gaining and maintaining synchronization.

The first method, shown in Fig. 4.32, is based on the transmission of a clock bit continuously at regular intervals; that is, a code synchronous bit must be transmitted for each character. The receiver is equipped with a free-running multivibrator which locks itself when it coincides with the extracted timing pulse information filtered out of the received data stream. To maintain constant accuracy the receiver clock timing must be frequently corrected, although the multivibrator remains locked even if it misses a number of transmitted synchronous timing pulses. The disadvantage of this system is that

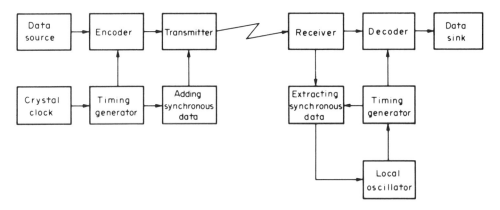

Fig. 4.31. Schematic block diagram of synchronous data transmission.

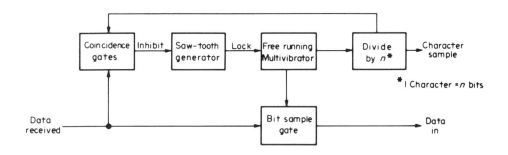

Fig. 4.32. Simplified block diagram of continuous character synchronization.

a relatively long acquisition time is required before synchronization is gained, and a number of 'empty' characters must be transmitted before the receiver establishes synchronization with the transmitter.

A better solution is to synchronize on the individual data bits received, thus omitting the need for the transmission of synchronous information bits with every character. This system, however, requires at least one bit transition from 1 to 0 or 0 to 1 to occur for each character transmitted. The start of the character is obtained once for each block by having a distinctive code transmitted at the head of the block which can be extracted by a suitable logic circuitry which resets all the counters to zero. This system is the most popular in use owing to the favourable feature that very little time is wasted in achieving synchronization.

Figure 4.33 shows a schematic diagram of a synchronization circuit where the clock timing bits are extracted from the received data. The correction must be performed slowly so that when synchronization is obtained it is maintained also in the absence of received bit transition. The bit period is sampled by a fast clock of N pulses, where $1/N$ shift is achieved for each bit transition. The clock timing is counted down to produce the accurate timing of a bit, and the internal bit of the receiver is then compared with the transmitted data bit. If the internal clock-timing pulse arrives before the data bit, then the internal pulse is advanced by one sample pulse; if it arrives after the data bit, then it is held for one sample pulse period. In the absence of an input data bit, it first advances by

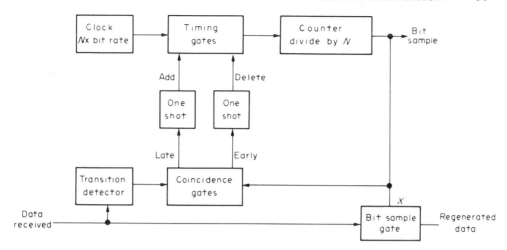

Fig. 4.33. Simplified block diagram of bit synchronization.

one sample at the beginning of the internal pulse and then deletes one sample at the end of the pulse, resulting in no change of the internal timing.

The synchronization technique described is based on the synchronization of the entire block with a unique code. Special codes are transmitted to establish formulating situations, which are extracted from the received data and thus recognized by the logic gating. This is performed by two or more synchronous characters at the beginning of each transmitted block. As a result, the synchronous information overhead is reduced to two characters per block, although the logical circuits involved increase the capital investment for each modem. Nevertheless, it is cost effective when considering the full utilization of the transmission path.

4.12 LINE DISTURBANCES

The discussion so far has concentrated on the design of the modems, without due regard to the problems of the transmission path. These now call for attention, for even with the best design of a modem the voice telephone line can cause distortion of the transmitted information. While the human ear is conditioned to tolerate most transmission disturbances, the telephone-line characteristics must be adapted for the acceptance of data transmission or to condition the data to accept possible errors. Noise on the line does not affect human conversation, although it could change the intelligence of transmitted data. The telephone line possesses the characteristics of varying amounts of amplitude and delay distortions which affect the transmission of the data. These distortions must be either tolerated or rectified. A telephone line could accept a limited amount of data without conditioning, although only for slow speeds. As the speed increases, the required bandwidth is widened, increasing the susceptibility to noise and other distortions. Line conditioning can be provided by adding special circuitry that reduces the line disturbances. In turn, data conditioning can be achieved by adding redundant data for error detection and correction.

The line disturbances can be classed in three groups, as follows:

(a)   Distortion due to the line characteristics.

 1. Delay distortion.
 2. Attenuation distortion.
(b) Noise induced on the line.
 1. White noise.
 2. Impulse noise.
 3. Phase jitter noise.
 4. Cross-talk noise.
(c) Equipment on the lines.
 1. Filters.
 2. Companders.
 3. Echo suppressors.
 4. Trunk signalling.
 5. FDM carrier shift.

The various line disturbances are discussed below. The graphs presented must be regarded as examples only, as they vary from line to line.

Delay distortion

When transmitting a spectrum of frequencies, some of the frequency components are delayed more than others. This phenomenon, shown in Fig. 4.34, is also termed "envelope delay distortion", since it destroys the pattern envelope of the frequency transmitted, and can cause differential delays between the various frequencies comprising the composite spectrum of the signal transmitted. This condition is manifested by some frequency components which are delayed much more than others, and which tend to become quite appreciable at both the lower and upper edges of the frequency band. This distortion cannot be detected by the human ear or be conveniently measured with common test equipment, though it can be measured with suitable test equipment. It could, however, create problems in data transmission, because all the frequency components do not travel through the transmission path at the same speed. These differences could cause an increase intersymbol interference where the data signal may spill over and disturb an adjacent data signal. The envelope delay distortion is the result of impedance mismatch between the line sections, primarily by the inductive and capacitive reactance elements in the transmission path.

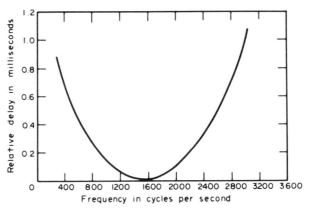

Fig. 4.34. Typical envelope delay for a voice telephone line.

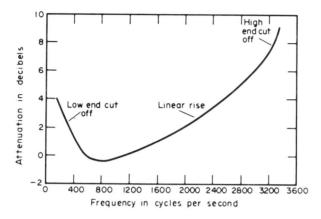

Fig. 4.35. Typical attenuation characteristics of a voice telephone line.

Attenuation distortion

The frequency bandwidth of the telephone line represents a varying frequency attenuation. This effect generally appears at the band edges, where there is a fast roll-off which results in larger frequency attenuation (as shown in Fig. 4.35). This is usually caused by filters in the transmission network path which consequently reduce the net of the available bandwidth in practice to about half that of the voice telephone spectrum, i.e. about 1.5 kc/s, unless special treatment is provided. The effect of the narrowing of the bandwidth will create transients in the data when it is transmitted at high speed. Attenuation distortion is less a problem in voice communication than in data transmission owing to the redundant nature and low information content of speech. In data transmission any alteration or loss of the symbol transmitted changes the meaning of the transmitted information.

White noise

White noise causes a disturbance which is uniformly spread throughout the frequency band and is occasionally noticeable on voice channels as a background hiss. It is composed largely of the thermal noise of the transmission path which interferes with the speech much more than with data. With data transmission the signal threshold helps to prevent this noise from affecting the data reconstruction. The white noise is always present and widely scattered. It is usually about 30–40 dB below the received signal level, though it may cause individual bit changes when the signal received is very weak. The random errors may be detected by special logic circuits.

Impulse noise

Impulse noise causes the main troubles in data transmission. It takes the form of sharp spikes which generally come in bursts of energy arising from switching and signalling equipment in the telephone network. These impulse noises usually do not critically affect human conversation but can have rather destructive effects on the transmitted data. The only remedy for impulse noise is to transmit the data in a block which includes redundancy data, using error-detecting codes so that all the errors are detected, and then have the facility to request the retransmission of the block when an error is detected It should be noted that errors caused by impulse noise tend to cluster in groups.

Phase jitter noise

A Phase Jitter noise is a random jitter which may be caused by the high speed of transmission where too few carrier cycles per bit are transmitted. It may also be caused by harmonics of the power-supply frequency which may change the phase of the carrier frequency. Another important cause of phase jitter is the modulation switching of the carrier when not in phase. The latter could be quite considerable in frequency shift. The phase jitter might cause quick phase shifts in the frequency in excess of 15-20°. This could reduce the speed of transmission when multi-state binary is used. Due to the characteristics of the equalizers, it has been found that phase jitter affects the modems with equalizers more than those with no equalizers.

Cross-talk noise

Cross-talk is caused by coupling between adjacent lines in cables where one line picks up signals travelling on another line. Any line running parallel with another line is bound to have some cross-talk coupling between them. In theory the lines are suitably balanced and cross wired to prevent the coupling, but in practice cross-talk is unavoidable. Cross-talk is affected by the induction between lines which increases as the line length is increased.

Cross-talk could also occur between adjacent channels in a frequency division multiplex where there are a number of different channels transmitted together. Even when the guard spacing is placed between the channels, some cross-talk is unavoidable. Another cause of cross-talk could occur with the leaking through of the supervisory control signal in a trunk channel.

Cross-talk is easily detected by the human ear, which is conditioned to weak signals. However, since the amplitude of the cross-talk noise is barely greater than that of the white noise, it causes negligible errors to the data transmission. It is interesting to find that a noisy cross-talk line where conversation is barely possible could be used successfully for transmission of 4800 bits/s.

Filters

Most telephone lines have a number of filters inserted along them. Nearly all the repeaters have some form of filtering which reduces the transmitted bandwidth. The utilization of the transmission path along a carrier system immediately calls for a number of filters, the effects of which cause envelope delay, as has already been discussed.

Companders

Many trunk lines which pass through frequency-division multiplex systems are designed with special equipment which compresses the loud speech signals and expands the weak ones. The special equipment is referred to briefly as companders. The companders are needed to ensure that the transmitted speech channel will efficiently utilize the multiplex line and not spill over into adjacent channels.

The companders are not required for data-transmission systems, as the data signals are generally transmitted at a relatively constant output level. It is advisable to avoid the use of trunk lines which possess companders and to select lines specially suitable for data transmission, but this may be unavoidable in many cases.

Echo suppressors

On some long trunk lines, equipment is inserted to avoid the return of the echo to the sender. The equipment is known as an echo suppressor and is based on the principle of preventing transmission in both directions simultaneously. It is used in some long-distance speech communication lines and will not cause interference there, since speech conversation is generally in a single direction at any given time. The echo suppressor is designed to ensure that the strong signal dominates and suppresses the weak signal. It takes about 120 ms to reverse the direction of the voice transmission, which is not noticeable for speech but could be critical in data transmission. Although echo suppressors are not needed on modern trunk transmission, they are still found in old audio circuits working on loaded cables.

Echo suppressors have two drawbacks if used for data transmission. The first is that they could prevent the transmission of data in both directions at the same time. This means that a line which possesses this equipment could not be used either for duplex transmission or for backward supervisory control. The second is that the use with half duplex transmission will require the addition of an artificial delay to prevent loss of data after the transmission reverses direction.

When hiring dedicated private lines care must be taken to ensure that no echo suppressors are inserted along the line. In dialled public lines this could be unavoidable and here, where such a line is detected, the only suggestion that can be given is to close down and redial again.

Trunk signalling

Many trunk signalling devices employ special control signalling on certain frequencies within the voice bandwidth. These discrete frequencies are filtered out at the receiver end in a way that does not affect the intelligibility of the voice transmission. Data transmission, however, could be disturbed as the frequencies of the spectrum near the vicinity control frequencies are attenuated. In practice this effect is not critical, as most of the data transmission spectrum is outside the range of the control signals even when some of the modulation signals components have some sidebands within the restricted range. There is a danger, however, that these sidebands may cause a false operation of the trunk signalling equipment in the telephone exchange.

Figure 4.36 illustrates the power-level restriction for data transmission over switched connections in the United Kingdom. The trunk signalling equipment transmits frequencies in the ranges marked as A and B. In range A, frequency signal components up

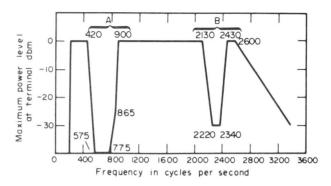

Fig. 4.36. Power level restriction for data transmission over switched connection. (Redrawn from *POEEJ*, London, Oct. 1966, by M.B. Williams)

to 0 dbm may be permitted if accompanied by signals in the area of 1 to 2 kc/s. In range B, signals are permitted if their characteristics preclude false operation of the trunk signalling equipment. In other words, the data modulated waveform must avoid the transmission of sidebands in this range for periods long enough to operate the signalling equipment.

FDM carrier shift

In the voice transmission using the frequency division multiplex carrier system, the voice bandwidth must be modulated so that it can be transmitted over the required channel bandwidth. The carrier frequencies of the modulator and demodulator may not be identical and will then cause some phase shift between them. This may not be recognized by the human ear but may still affect the correct interpretation of the recovered data.

Compensating the disturbances

All the disturbances discussed above could be disregarded by the human ear when speech is transmitted although for data-transmission applications it must be compensated. Noise disturbances can be compensated by adding redundant data which enables the detection of the errors caused by the noise. The error-detection technique is not part of the transmission media and will therefore be discussed separately (in Chapter 12). The present chapter deals with the analogue representation of transmitted digital information, that is, it is mainly concerned with the modem, leaving the logical handling of the error detection to be dealt with as a separate topic. However, as the distortion of the frequency spectrum has a direct effect on the analogue part of the data transmission, the modem itself must provide means of compensating any distortions caused by the line mismatching.

The delay and attenuation distortions can be reduced by adding delays to selected portions of the transmission frequency band. Both amplitude and delay distortions create intersymbol interference which limits the speed of transmission. These distortions may be either tolerated, accepting the speed limitations, or equalized to allow higher speeds, a

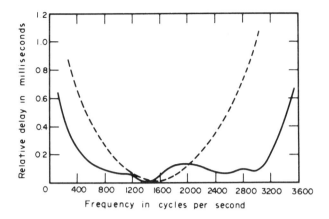

Fig. 4.37. Typical equalization effect on compensating envelope delay.

process which will add to the cost of the transmission of each bit.

Envelope delay distortion can be reduced by adding delay elements to selected portions of the frequency spectrum, thus providing an ideal relatively uniform delay, as shown by the graph given in Fig. 4.37. The same approach could be used for the attenuation distortion where the equalizer adjusts to compensate the roll-off spectrum characteristics.

There are three types of equalizers which may be used: fixed, manual adjustable and automatic adjustable.

Fixed compensated equalizers

A common approach to equalizer design is to use a fixed matching network based on the known line characteristics. This will improve the theoretical line at the critical points of the frequency band. It centres the compensated equalization of variation of the distortion at about zero. The design is based on defining the ideal line characteristics and then theoretically compensating it. This has many practical limitations and may not be suitable for all the lines. For modems operating with speeds up to 1200 bits/s, it is usually adequate to use fixed equalizers. The fixed compensated equalizer provides nominal improvement of very bad lines; however, their use will sometimes impair, rather than improve, an existing good transmission line, since it adds attenuation to the line.

Manual adjustable equalizer

This system is based on a set of fixed equalizers which permit the matching of the transmission line to a specific equalizer network. This adjustable equalizer can compensate for a wide range of line characteristics. Its installation requires skilled personnel and special measurement test equipment to match each transmission line. Admittedly the process of installation is slow and far from perfect. Experience shows that the equalizers are installed rather by the technician's intuition than by his skill. Each time the modem is switched to a new line it ought to be rematched to the line, a practice which is seldom observed. The manual adjustable equalizers are used in modems with

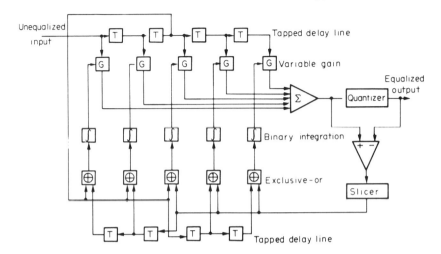

Fig. 4.38. General configuration of an automatically adaptive equalizer.

speeds up to 2400 bits/s. In these modems there is a set of fixed compensating equalizers which can be chosen to match a range of expected line distortions.

Manual adjustable equalizers allow the modem to be adapted to operate over a number of lines, but they will still be limited in multi-level data transmission where the intersymbol interferences might be too great, making it difficult to correct with fixed equalizers.

Automatically adjustable equalizers

Both the fixed and the manually adjustable equalizers are limited in use, since they are restricted to the ideal transmission line. The future trend is towards automatically adjustable equalizers which adapt themselves to the specific line being used, thus eliminating the need for any manual adjustments. For higher speeds, where multi-level codes must be used, the fixed and manual adjustable equalizers cannot provide rapid correction of the distortion. Where the number of states is more than two, the intersymbol interference could be so severe that it would limit the speed. The equalizer line matching must therefore be rapid and maintain continuous correction.

Figure 4.38 illustrates a general configuration of an automatically adaptive equalizer. The demodulated signals pass through a delay line, having taps at the symbol intervals. The time delay of each delay stage is equal to T, which is the Nyquist interval time. Each tap is provided with a variable gain G and the various outputs are summed up to provide the equalizer output. In the feedback control circuit the equalizer output and the delay tap are each delayed relatively to one another by shift registers. The outputs of the shift registers undergo binary multiplication in exclusive-OR circuits. The binary outputs are fed to binary averaging circuits before being used to control the variable-gain circuits.

4.13 HANDSHAKING BETWEEN MODEMS

The key factor in data transmission is the transmission channels which permit the 'handshaking' of two modems. The designer has many options of transmission channels but in most cases they are restricted to the available voice telephone line. There are many other options, such as radio channels and wide band cables, but these are outside the immediate discussion. When using telephone lines there are two choices, dialled or leased lines. Both these types are generally the responsibility of the communication Common Carrier Company. In the case of dialled lines, the cost varies with the distance and operating time while in the case of leased lines the cost is paid on a monthly basis which is according to the type of line and its distance. Dialled lines are referred to also as public lines while leased lines are referred to also as dedicated lines (since they are generally directly attached to the computer and the remote terminal). Another type of telephone lines are private lines which are installed by the user and do not pass through the public network.

In direct distant dialling (abbreviated DDD) no special transmission line is available, because the lines are obtained randomly. The user has no choice of line performance and must be satisfied with whatever is available. Sometimes the line obtained is a relatively good one, but normally the line obtained is noisy, with impaired characteristics. It is essential to regard each line with caution, checking whether its characteristics are suitable. It may have equipment attached to it which would not only disturb the transmission but may even prevent it. The dialled lines usually carry control signals within the bandwidth and/or other equipment required for voice communication, such as filters, echo suppressors, etc.

The main advantages of a dialled line is that a connection can always be readily found, and so it can operate most of the time. Furthermore, it is not restricted to any point-to-point connection but can be routed and connected to any location site anywhere in the world provided it is linked to the public dialling telephone system. In dialled lines there is no need to prepare for alternative lines, as this will be provided by the exchange.

With today's technology (of equalizers) there are modems available which are said to operate up to 4800 bits/s on dialled lines. The author would not, however, recommend the use of the public telephone lines for transmission speeds above 2400 or even 1200 bits/s. This is because of the possibility of impaired randomly accessible lines which may be encountered. When it is required to transmit with speeds of 4800 bits/s on dialled lines the modem will require good error-correcting equipment and automatically adjustable equalizers. Even then it is doubtful whether this speed will be able to be transmitted on all the calls dialled. On average it will prove to be wasteful, if the time lost in seeking to make the call on a suitable channel is prolonged. The cost trade off of the extra equipment and the delay in obtaining the proper dialled contact makes the attempt to achieve higher speeds questionable. With today's technology, dialled connection is quite adequate for most low-speed applications. If higher speeds are required the only practical solution is the use of dedicated lines.

For most Command and Control applications it is advisable to utilize the dedicated, leased or private lines, disregarding the speed of transmission, since the main criterion in Command and Control systems is the design of real-time and this cannot be obtained when dialled lines are used. With the use of leased or private lines, a reliable channel is ensured, because the lines are specially selected. Furthermore, the dedicated line is always available, as it is permanently connected to the location site and its characteristic performance is known. Once the line is reserved, it may be separately compensated for distortion.

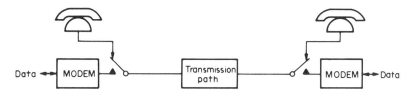

Fig. 4.39. Modem handshaking with speech facilities.

Dedicated lines are generally of high quality and present a low error rate. However, leased lines may be channelled, without the user's knowledge, over carrier cables or microwave radio links. In addition, although theoretically permanently connected, changes are sometimes made to the routing without their being reported to the user.

There are generally three types of dedicated telephone links which may be leased, 4A (C1), 4B (C2) and 4C (C4), where the options required depend on the speed, i.e. the bandwidth available. Table 4.4 shows the general characteristics of these voice lines.

Leased lines cost more than dialled lines and the price doubles if alternative lines are also made available. When a leased line breaks down it takes time to locate the fault or to obtain an alternative line if one was not provided for originally. Leased lines are thus much more expensive than the dialled lines, although the cost effectiveness is reversed when there is a large amount of data to transfer. The break-even point depends on the actual charges for the various lines and the amount of data transmitted.

Establishing a connection between two points, a process which is known as handshaking, depends on whether the line is leased or dialled. In dialled lines the connection is first made by dialling the required location (as can be seen from Fig. 4.39). After preliminary conversation, both ends switch over the key situated on the modems

TABLE 4.4

Condition	4A (C1)		4B (C2)		4C (C4)	
Frequency stability	±5 c/s		±5 c/s		±5 c/s	
	Spread limits	Frequency range (c/s)	Spread limits (db)	Frequency range (c/s)	Spread limits (db)	Frequency range (c/s)
Amplitude distortion	−2 – +6 db	300-999	−2 – +6 db	300-499	−2 – +6 db	300-499
	−1 – +3 db	1000-2400	−1 – +3 db	500-2800	−1 – +3 db	500-3000
	−2 – +6 db	2401-2700	−2 – +6	2801-3000	−2 – +6 db	3001-3200
	Overall delay	Frequency range (c/s)	Overall delay	Frequency range (c/s)	Overall delay	Frequency range (c/s)
Envelope delay relative to 1000 c/s	1000 (ms)	1000-2400	500 (ms)	1000-2600	300 (ms)	1000-2600
			1500 (ms)	600-2600	500 (ms)	800-2800
			3000 (ms)	500-2800	1500 (ms)	600-2000
					3000 (ms)	500-3000

from speech to data modems. The modems at both ends must then first establish synchronization before data can be transmitted.

There are dialled lines where one of the modems is connected permanently to the lines while all the other modems may be manually operated. This mode could be arranged at the computer end which distributes its services to a number of customers, each having a terminal. In this system the operator dials the fixed station and if it is free the operator can then hear a unique return tone. He can then transfer his modem to data and transmit his message. In this application he must first transmit a special code so that the computer can recognize the caller.

In leased lines the connections may be fully automatic, where the remote terminals are permanently attached either to the central processor or to the central control via the modems and lines. In this state, the central processor 'master' controls the 'slave' terminals, that is, one side controls the data flow in the lines by asking each terminal in turn whether it has any data to transmit. The modems in this network are 'transparent' and all the handshaking is controlled automatically without human intervention.

In some leased lines there could be special telephone instruments associated with the modem which permit calling the other end by means of a ring tone. These telephone sets cannot be the same as the standard post office instrument, since in leased lines there is no power and the lower frequency band is restricted. (The low frequency is cut off when the data is transmitted over carrier cables or radio links.) In dialling lines, a special channel of the carrier equipment is dedicated for the signalling; this, however, is not feasible in leased lines, as each line is operated separately. Furthermore, magneto telephones, which may be operated on wire leased lines, will not be able to operate on these lines because the signalling frequency of the magneto telephones is very low. For this reason modems which operate on leased and private lines require a specially designed signalling if a telephone is to be used on the lines. In these modems, a particular tone is transmitted from one modem which is recognized by the other modem even when it is switched to the data mode. In the author's opinion all modems should be equipped with the option of telephone speech facilities, even if all the connections are made automatically by the computer. This facility is mainly intended for the benefit of the technicians and not for the operators, since it is important essentially for servicing and maintaining the system operation. In the long run the initial extra cost of the speech facility pays, if it enables a technical or an operational fault to be located without delay.

4.14 CONCLUDING COMMENTS

In the introduction to this chapter it was stated that the essence of Command and Control systems is the means of transferring data between remote geographical locations. Nevertheless, when designing the data transmission configuration it must be kept in mind that data transmission is not a target on its own, but is a service of the Command and Control system.

The main factor to be considered in the adoption of any system is its cost, or rather its cost effectiveness. In data transmission (although not always in Command and Control) the cost refers to the price of transmitting a bit of information. The cost of the modem, which is the most complex equipment in data transmission, accounts for only a small part of the overall cost required for the transmission of a bit of information. Here the modem selected should be judged not by its price alone but mainly by its performance in exploiting the line. An important consideration in the overall cost is the cost of the transmission line itself. In specific cases it pays to take the most expensive modem if it ensures the most efficient utilization of the transmission line. This would not be applicable if the lines are not to be utilized to their full capacity.

In Command and Control systems which operate in a real-time on-line mode, the designer should be aware of when he is dealing with the cost of the transmission of each bit. This is an average statistical figure, generally measured over a period, say, 24, 12 or 6 hours. In Command and Control systems there may be an extreme application where there is a need to transmit only a few messages per day, but if these messages are required urgently (i.e. in real-time) it is important then to reconsider the cost methods in use. Judging the cost of the transmission media should be based on the cost of transmitting a bit with a delay of N seconds. This is necessary in order to present it in real-time environment. The volume of data plays but a small part in calculating the speed where the essential part is the delay in transferring the volume at the peak period. The cost effectiveness here should not be based on any average calculations but on the price the designer is ready to pay to achieve the so-called 'real-time'. The designer must compromise between the price of transmitting a bit at the peak period and the delay time required to achieve real-time.

The designer is confronted with a vast range of modems. Table 4.5 shows some of the

TABLE 4.5

Speed	Channel	Trans. line	Modulation	Sync.	Comments
75 150 200 300 600	Very low speed Sub-voice lines	Part of a voice line	AM FSK	 ASY	Used in parallel with voice or other lines
600 1200	Low speed Voice lines	Dialed lines, 4A	FSK	ASY/SYN	
1200 1800 2400	Medium speed Voice lines	Leased lines, 4B	FSK PSK	 ASY/SYN	
2400 3600 4800	High speed Voice lines	Leased lines, 4C	Duobinary DPSK VSM	 SYN	
4800 5400 7200 9600	Very high speed Voice lines	Conditioned lines	VSM DPSK QAM	 SYN	Limited distance if not properly equalized
50K to 20M	Ultra high speed Wide band lines	Leased or private Wideband lines T1	PCM VSM DPSK QAM	 SYN	Operates with multiplexors

modems available.

A side aspect of the modem specification is defining how it will be connected to the line. Some modems are hard-wired to the line while others are only acoustically connected. Acoustically coupled modems are used on dialled lines and have the advantage that they can be attached to any public telephone set (though limited to 1200 bits/s). The direct hard wired is preferable, as it is more secure and error free. In Command and Control systems there is no room for acoustically connected modems.

Many telephone companies provide the modems with the transmission lines so as to safeguard their equipment. Some only provide the interface unit between the lines and then the modem must be designed to protect the telephone line from electrical disturbances and from overloading it. (The maximum power output of the modem must not exceed 1 mW.)

Although the physical shape and the installation have no bearing on the modem's performance, they do affect the selection of the modem. The size and shape of the modem are obviously important factors, since room must be found to install the modem. For Command and Control systems, where many modems may be required to be situated at a single installation, a further aspect should be considered. Here, it is preferable to mount the modems in a single rack using plug-in boards for the modems. This will allow the installation of a common power supply and even a common clock source, which will reduce both the space and the cost.

Most modern modems are provided with self-test circuitry to aid in locating faults. If a large number of modems are assembled together extra test equipment could be added to permit the testing of the line characteristics.

4.15 REFERENCES

1. Bennett, W.R. and Davey, J.R., *Data Transmission*, McGraw Hill Inc., 1965.
2. Lucky, R.W., Salz, J. and Weldon, E.J., *Principle of Data Transmission*, McGraw Hill Inc., 1968.
3. Richard, R.K., *Electronic Digital Systems*, John Wiley & Sons, 1966, pp. 362-450.
4. Martin, J., *Telecommunication and the Computer*, Prentice Hall Inc., 1969, pp. 239-364.
5. Nyquist, H., "Certain topics in telegraph transmission theory", *AIEE Trans.*, Vol. 47, pp. 617-644 (Apr. 1928).
6. Karnaught, M., "Three level binary code transmission", U.S. Patent 3 214 749, Oct. 26, 1965.
7. Alexander, A.A., Gryb, R.M. and Nast, D.W., "Capabilities of the telephone network for data transmission", *Bell System Technical Journal*, Vol. 19, No. 3 (May 1960), pp. 255-286.
8. James, R.T., "Data transmission – the art of moving information", *IEEE Spectrum* (Jan. 1965), pp. 56-83.
9. Pierce, J.R., "Some practical aspects of digital transmission", *IEEE Spectrum* (Nov. 1968), pp. 63-70.
10. Lender, A., "Correlative level coding for binary-data transmission", *IEEE Spectrum* (Feb. 1966), pp. 104-115.
11. Franklin, R.H. and Law, H.B., "Trends in digital communication", *IEEE Spectrum* (Nov. 1966), pp. 52-58.
12. Hersch, P., "Data communication", *IEEE Spectrum* (Feb. 1971).
13. Davey, J.R., 'Modems', *Proc. IEEE*, Vol. 60, No. 11 (Nov. 1972), pp. 1284-1292.
14. Hirsch, D. and Wolf, W.J., "A simple adaptive equalizer for efficient data transmission", *IEEE Trans. Comm. Tech.*, Vol. COM-18, No. 1 (Feb. 1970), pp. 5-11.
15. Westcott, R.J., "An experimental adaptive equalized modem for data transmission and the switched telephone network", *The Radio and Electronic Engineer*, Vol. 42, No. 11 (Nov. 1972), pp. 499-512.
16. Franklin, R.H. and Rhodes, J., "Data transmission", *Journal IEE* (Feb. 1962), pp. 82-85.
17. Williams, M.B., "The characteristics of telephone circuits in relation to data transmission", *The P.O. Electrical Engineers' Journal*, Vol. 59, Pt. 3 (Oct. 1966), pp. 151-162.

18. Wier, J.M., "Digital data communication techniques", *Proc. IRE* (Jan. 1961), pp. 196-209.
19. Roberts, L.W. and Smith, N.G., "A modem for the datal 600 service", *P.O. Elect. Engrs. J.*, Vol. 59, Pt. 1 (Apr. 1966), pp. 108-116.
20. Muth, R.J., Shipman, J.E. and Toffler, J.E., "Factors affecting choice of modulation techniques for data transmission", *Telecommunications* (Mar. 1969), pp. 17-22.
21. Amoroso, F., "Bandwidth-efficient modulation for voice band data transmission", *Telecommunication* (Apr. 1968), pp. 19-24.
22. Murphy, D.E., "Introduction to data communication", *Telecommunication* (May 1968), pp. 23-30.
23. Chaney, W.G., "The future for data transmission on communication circuits", *Telecommunication* (Dec. 1968), pp. 9-16.
24. Murphy, D., "Digital techniques in data communication", *Telecommunication* (Feb. 1969), pp. 19-25.
25. Andrews, H.R., "Let's take the mystery out of modems", *The Electronic Engineer* (July 1972), pp. 5-10.

CHAPTER 5

Multiplexors and Concentrators

5.1 INTRODUCTION

The discussion in the previous chapter was centred on the efficient utilization of the data-transmission path of point to point communication. If the volume of data originating from any point is relatively small, no design planning can reduce the cost of each bit transmitted along the transmission path beyond a certain figure. The only way for further reducing the cost per bit transmitted is to find means of increasing the volume of data transmitted in the path, that is, sharing the same transmission path with other data sources. In this way the cost of the path is shared, and the cost of the actual bit transmitted is reduced. The designer should aim at operating a number of terminals together, sharing a single transmission line by using either multiplexing or concentrating techniques.

The principle of sharing the same line between a number of terminals is shown in Fig. 5.1. The multiplexor or the concentrator combines several different signals operating at low speed, sending them simultaneously over the same transmission path and then reconstructing them again at the receiving end.

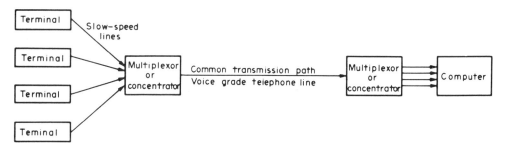

Fig. 5.1. Common transmission path shared between number of terminal.

The terms multiplexors, concentrators and also message switching are not always clearly defined in the technical literature and in many cases may be variously applied. Communication multiplexors generally refer to the direct utilization of the transmission path by the sharing of the resources of time or frequency. In other words, the sharing is based on the static allocation of the resources by means of a fixed predetermined method. The concentrators also share the resources, but in contrast to multiplexors, they utilize the line dynamically. In this case, the line resources are shared randomly and not according to a fixed arrangement. The data generally reaches the line at a staggered rate, which is not important when multiplexors are used, as each input there receives a fixed part of the line irrespective of the traffic volume. On the other hand, with concentrators better efficiency can be obtained. The function of the concentrator is to smooth the data flow in the transmission path, aiming at the continuous transmission at the maximum possible speed.

Multiplexors were discussed in Chapter 3 but there the consideration was limited to telemetry application, while here it is concentrated on digital data.

5.2 FREQUENCY-DIVISION MULTIPLEXORS

Frequency-division multiplexors share the frequency spectrum of the transmission path among a number of data channels. Each data channel receives a unique frequency band which is permanently allocated to the channel. If the full bandwidth F is divided into N channels, then each channel has the frequency bandwidth of F/N. However, each channel can transmit at speeds far less than the frequency slot of F/N available to it. The limitation is due to the need of guard bands between adjacent channels. The guard frequency bands prevent any sideband signals from overlapping and encroaching on the adjacent channels. Although bandpass filters are used at the output of each channel to envelop the allocated frequency band, the slope of the filters must be within the range of the guard bands. When voice grade lines are used with frequency-division multiplexors, only a total maximum speed of about 2000 bits/s can be reached.

In frequency-division multiplexing the transmission of the data in all the channels is in parallel form. That is, the transmission path can transmit different information over each channel simultaneously. Alternatively, each channel can transmit a bit belonging to the

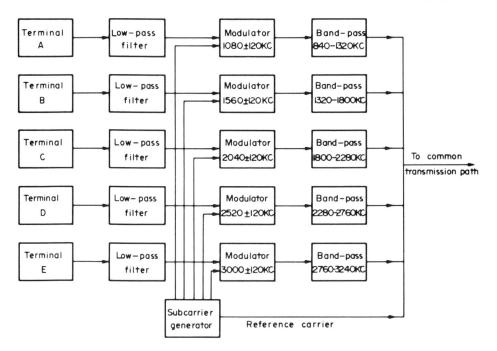

Fig. 5.2. Typical frequency-division multiplexing scheme (the common voice grade path is divided between five channels, each transmitting at a maximum speed 200 bits per second).

same character, thus transmitting the character in parallel. Frequency-division multiplexors may be used to share a one voice grade line among a number of slow-speed terminals, or alternatively may be used to share a wide-band cable among a number of voice channels. The latter has been in use for many years in carrier trunk systems.

The frequency translation between the original bandwidth and the frequency bandwidth allocated to each channel is performed with modulating techniques (as shown in Fig. 5.2). The modulation techniques used are in general amplitude or frequency modulation. The frequency-division multiplexor system comprises a number of parallel channels each consisting of filters and a modulator. Because each channel modulator is a separate independent modem transmitter unit, the failure of one will not affect the other adjacent modulators.

The output from the terminals passes through a low pass filter before the signal enters the modulator. The function of this filter is to smooth the sharp edges of the data pulses (as was explained in the previous chapter). The output of the modulator is passed through a bandpass filter in order to prevent any side bands from disturbing the adjacent channels. The frequency modulated signals from all the channels are fed simultaneously to the transmission path.

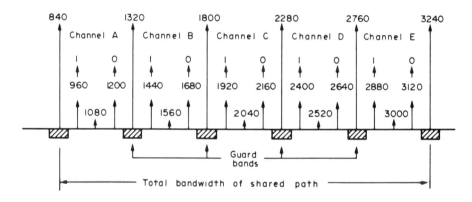

Fig. 5.3. Spectrum allocation of the transmission path bandwidth in a typical frequency-division multiplexing scheme.

The frequency spectrum of the complete voice transmission path bandwidth and its allocation to each individual channel is shown in Fig. 5.3. In the example (shown in Figs. 5.2 and 5.3) the voice grade telephone type line is shared between five channels. Each channel is allocated a bandwidth of 480 c/s and can transmit speeds up to 200 bits/s. The distance between the digital '1' bit and the digital '0' bit in each channel is 240 c/s leaving another 240 c/s spacing between the signals of adjacent channels. The band pass filters must filter the output of each of the five individual modulators without their interfering with each other. This necessitates restricting the bandwidth allocated to each channel to less than 480 c/s.

The only common equipment in the frequency multiplexor is the subcarrier generator, which ensures that the frequency bandwith allocated to each channel is related to the others and prevents individual shifts in the subcarrier frequency. For the same reason, a reference carrier is transmitted with the multiplexed signal, so that all the demodulators can be related to this reference.

5.3 TIME-DIVISION MULTIPLEXORS

In time-division multiplexors, the line spectrum is also shared between a number of data channels, although here each channel gets a different unique time slot. During the period of each time slot, the channel gets the full frequency bandwidth of the transmission line. The terminals are scanned in a fixed sequence, thus producing the data transferred to the line in a serial form. The function of the time-division multiplexing technique is now to transmit several simultaneous messages by interleaving the data samples from many sources.

The period of each time slot may be allocated to a bit, a character or a block. Bit interleaved time multiplexors have the advantage of simplicity and present minimum delay. Character interleaved time multiplexors are generally more expensive and present more propagation delay since they require the buffering of a complete character. Nevertheless, the character interleaved multiplexors permit more efficient utilization of the transmission line. Block interleaved multiplexors are seldom used in the simple TDM discussed here although widely used in other TDM applications, as will be shown later.

A typical time-division multiplexor scheme is shown in Fig. 5.4. With each clock time pulse the transmission line is presented to another terminal and remains attached to it for the full period between the clock pulses. This time slot period can be short and equal to a bit or be as long as required, but the period must always be of equal duration in all the channels.

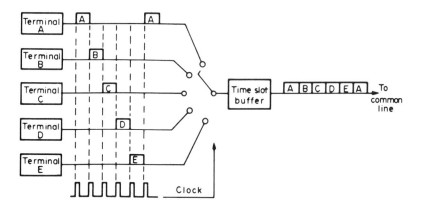

Fig. 5.4. Typical synchronous time-division multiplexing scheme.

Time-division multiplexing principles are widely used in many other fields. One of these applications worth mentioning is time-shared computer systems (to be discussed in Chapter 9).

Each channel in the time-division multiplexor is allotted a fixed permanent time slot. The serial data is framed in fixed formats and, in general, each frame is started by a unique bit sequence required for synchronization. It must be pointed out that time-division multiplexors can operate only in a synchronous mode, as both ends of the line must scan the terminals using the same synchronous reference. That is, they must scan the terminals at a specific sequence which must be identical both at the transmitting and receiving ends. For this reason these multiplexors have been termed Synchronous Time-division

Multiplexors (STDM) to distinguish them from Asynchronous Time-division Multiplexors (ATDM) which will be discussed in Section 5.6. Although the multiplexors and demultiplexors at both ends of the line must operate in synchronization, the data transmitted may be in the form of synchronous or asynchronous format. Each channel can utilize the time slot allotted to it according to the data originating from the terminal, and is independent, within the time slot period, of the way the multiplexor is operated.

Similarly to frequency division, time-division multiplexing also requires spacing between the channels. This guard spacing is generally used here for synchronization. In some applications, where bit interleaving is used, the synchronization indicates only the start of the scanning cycle. In character interleaving multiplexing schemes a synchronous bit is required between the time slots.

The main advantage of time-division multiplexors relative to frequency division multiplexors is the overall data transmission speed that can be reached. Since the speed is a direct function of the bandwidth, the overall speed reached in frequency-division multiplexors is limited to 2000 bits/s. In time-division multiplexors the data is transmitted in series using the full available frequency spectrum for each channel. Thus in modems associated with time-division multiplexors, the data transmitted could be arranged in multi-level form, transmitting two, three or four bits for each state. Hence, with time-division multiplexors, the associated modems may be operated at the maximum possible speed, reaching 4800, 7200 and even 9600 bits/s.

Another advantage of the time-division multiplexor is that only one modem is associated with the multiplexor, while with frequency multiplexors every channel must require a separate modem. It should be noted that the terminal in TDM must be close to the modems to prevent the distortion of the data pulses.

When a number of time-division multiplexors are required to operate in cascade, the signal on the line must be reconverted back to digital form before it can enter the TDM. That is, if the distance between two adjacent TDM's arranged in cascade is large, then an extra two modems must be placed on the transmission path, one at each end. Cascading time division multiplexors bring forward the problem of synchronizing all the scanners in the cascade. In general it is solved by having a master clock at one end of the cascade and synchronizing the whole network with it, using a slave-clock system in each consecutive TDM unit. Figure 5.5 illustrates a typical time-division multiplexor cascaded network.

A most important aspect of a time-division multiplexor is that it exhibits true digital form and does not require any analogue-to-digital conversion as in frequency multiplexing. This enables the data to be fed directly into the computer and also allows for further manipulation *en route* such as readdressing.

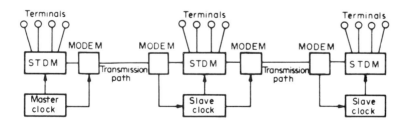

Fig. 5.5. Cascading synchronous time-division multiplexor units.

5.4 TIME-DIVISION MULTIPLEXORS VERSUS FREQUENCY-DIVISION MULTIPLEXORS

Both types of multiplexor schemes share the line resources, although in FDM all the channels operate in parallel, while in STDM the channel operates in serial each getting the full frequency channel. FDM is easier to implement, because with each channel can operate independently, yet STDM exploits the available transmission path to a greater extent. Figure 5.6 illustrates how the spectrum is shared between the channels in both multiplexing schemes.

Both frequency and time multiplex reduce the overall cost of the transmission

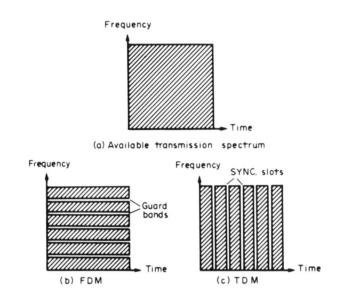

Fig. 5.6. Comparison between frequency and time-division multiplexing schemes.

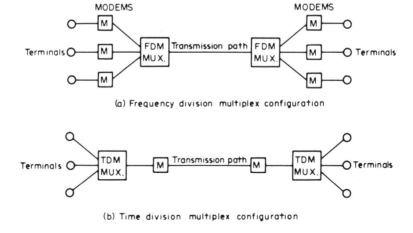

Fig. 5.7. The position of the multiplex and modem in the network.

network. The cost effectiveness of both systems depends on the distance between the terminals and the multiplexor. Synchronous time-division multiplex is more expensive than frequency-division multiplex, particularly when the remote terminals are not clustered at a single remote site. On the other hand, if many terminals are centred at one location, as is the case in many Command and Control applications, the STDM presents better cost effectiveness. This is because the line can be used to its maximum capabilities, that is, at its maximum transmitting speed.

In FDM the transmission speed is rather limited, while much higher speeds can be obtained in STDM. The expensive modem equipment required for the high speed is shared between the channels.

A most important aspect in the comparison of FDM and STDM schemes is the position of the modems in the network (as shown in Fig. 5.7). In frequency-division multiplexors the multiplexors are positioned on the line after the modem, while in time-division multiplexors the multiplexors are between the modems and the terminals. This demonstrates the cost effectiveness of both systems and their relation to the geographical spread of the terminals. The best economical network involving a diversity of terminal locations is the use of a hybrid cascade network of both FDM and STDM schemes, as shown in Fig. 5.8. The example here is for voice grade lines, whereas with wider band lines, more combinations are possible.

Another point of comparison is the form of signal that is multiplexed. In FDM the form is of analogue waves, which have the advantage of allowing the multiplexors to be at

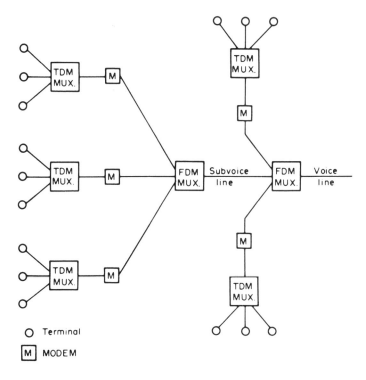

Fig. 5.8. Cascading both TDM and FDM in the same network.

TABLE 5.1

Function	FDM	STDM
Type of signal	Analogue waveform	Digital pulses
Type of terminals used	Used for low-speed terminals	Used for high-speed terminals
Geographical spread	Used for isolated terminals	Used for clustered terminals
Type of transmission	Par Variable to 2000 bits/s depen-	Serial transmission
Relation to the original signal	The full wave-continuous	Samples only
Spectrum limitation	Limited in frequency	Limited in time
Speed overall limitation	Variable to 200 bits/s depending on the bandwidth available	Variable to 9600 bits/s depending on the line and modem available
Position of multiplexor	On the line after the modem	Near the terminals before the modem
Data delay	No delay at all	One bit (or character) delay
Error-detection feasibility	No facilities possible	Feasible and practical
Independence of channels	Each channel operates separately and independently	The data in all the channels is interleaved
Switch facilities	Only in electromechanical or space switching schemes	Easy to manipulate on any electronic switching scheme, with no extra equipment required
Connection to the computer	Requires analogue-to-digital conversion before the computer can handle the data	Can be directly connected to the computer
Price of equipment	Relatively cheap	Expensive
Data redundancy	None	Special data for synchronous, framing and error detection
Guard spacing between channels	Frequency spacing	None beside synchronous information
Utilization of the line	The full effectiveness cannot be achieved	Efficient utilization of the full spectrum
Reliability	Susceptible to noise and distortion	Less affected by noise
Cascading	Readily possible	Requires the addition of a modem pair to each multiplex and a master–slave clock scheme

any distance from the terminal and be easily cascaded. IN STDM the signal is in the form of digital pulses, which have the advantage that they can be easily manipulated by any logical circuitry.

Table 5.1 shows some of the points of comparison between the two multiplexing schemes.

The main disadvantage of both FDM and STDM schemes is that the maximum number of terminals is limited for each multiplexor. A second relevant disadvantage is that each channel may not be fully used if there is no data to transmit from the associated terminal. In other words, the terminal gets the share allocated to it regardless of whether or not it has data to be transmitted. The cost effectiveness of the system is based on the maximum loading of all the channels. If the data originating from the terminals is much below the maximum, this type of multiplexing scheme may prove most inefficient.

The common principle of these two multiplexing schemes is that each terminal always has a transmission channel available to it, since a shared part of the line is fully confined to each terminal. For this reason the terminal can never find the line busy, as may occur in other schemes. Although this may be regarded as an advantage, it could conversely be regarded as a disadvantage, since the line is not always efficiently utilized.

Despite the disadvantages of both frequency-division multiplexors and time-division multiplexors, they are both widely used, because they reduce the overall cost of the transmission network. The most efficient utilization of these schemes is an integrated blend of both FDM and STDM, as has already been discussed.

5.5 THE CONCEPT OF THE CONCENTRATOR

There are many instances where the multiplexing schemes described do not fully satisfy the large system requirements, and consequently there is a demand for a more versatile data facility which could utilize the common transmission path more efficiently. This need has been met by introducing the concentrator, which is qualified to cater for a wide range of requirements. The term 'concentrator' was first applied to telephone switching systems, where economy was effected by combining many speech telephone circuits in a single channel.

In the previous sections the discussion was concerned with the efficient transmission of the data over a common path utilizing the multiplexors. As long as the data flow in the common path is continuous, the conventional multiplexing schemes discussed prove to be extremely satisfactory in operation. Degradation of the system efficiency is only apparent when the flow of the data is random.

The main disadvantage of the described multiplexors is that the number of terminals attached to a specific multiplexor is limited. Each channel in the common path is confined to a particular terminal, that is, each terminal dominates part of the common path in either frequency or in time range. The actual data transmitted on the multiplexed line has no relation to the maximum volume of data that can be transmitted, the full volume of which could only be achieved if all the terminals transmitted their data into the associated channels continuously at the maximum speed. This ideal situation almost never occurs, and in most cases the volume of data transmitted is at the minimum level. Both time and frequency division multiplexor schemes are designed to serve for the peak loads presented by the terminals and not for the nominal load.

It must be asserted that the multiplexor schemes as described are not practical where a large number of terminals are involved, for it would be extremely wasteful to incur the cost of the expensive equipment required for multiplexing if the terminals are not frequently involved.

In order to increase the efficiency of utilizing the common path, the volume of data must be strengthened. The only means of doing so (since the volume of data originating from each terminal is generally random and low) is by increasing the number of terminals attached to each common line. In view of the fact that the number of terminals attached to each multiplexor is limited, special techniques such as concentrations (which is a general term for a diversity of equipments and techniques) are required. The concentrator, as distinct from multiplexors, can provide the facility for each terminal to transmit at the maximum line speed, though some delay may be incurred.

Concentrators are distinguishable from multiplexors by the way they share the resources of the common path. Concentrators assign the line dynamically on a request basis, while multiplexors do so rigidly on a fixed basis. Another important distinctive feature is that in concentrators the common bandwidth is smaller than the sum of all the input bandwidths.

A further point must be noted. Concentrators are more flexible than multiplexors, with the number of terminals not limited by the line bandwidth but by the concentrator itself, that is, the number will depend on its operating power and its available buffering capacity. The multiplexors, on the other hand, provide a relatively economical solution to small-scale general communication problems. A concentrator, too, offers an economical scheme for the communication network, but it is intended essentially for large-scale networks where the traffic flow is considerable. In fact, the concentrator can operate as a multiplexor, since it is most flexible, whereas the multiplexor cannot act as a concentrator, since its operational functions are static.

The most simple concentrator is one which presents the line to each terminal upon request. In other words, the concentrator could operate as a simple fast switch that connects the terminal to the line upon request. Each terminal which has a message to transmit may request the line to transmit the message and release the line after transmission. If two terminals request the common line at the same time, then one of them must wait till the other has completed its transmission. In order to prevent the terminal from dominating the concentrator, the size of the message from each terminal must be limited. This concentrating scheme will require the terminals (in this case teletypewriters) to be locked whenever the line is engaged by one of the other terminals and a pilot light to be provided on each terminal to indicate that the line is engaged (as shown in Fig. 5.9). This scheme has very little advantage over the multiplexor and is used only in small scale low speed communication networks.

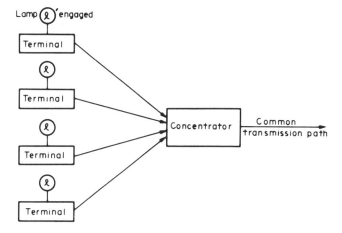

Fig. 5.9. Principle of a simple concentrator where each terminal can request the line. If the line is engaged the terminal is locked with the lamp indicating.

Another solution to the concentrator design problem is to have a buffer situated near each terminal which can hold the transmitted character until the common line is available

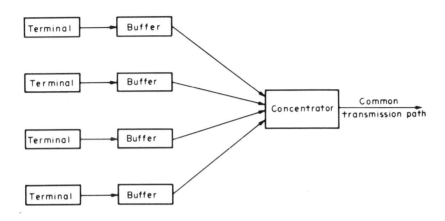

Fig. 5.10. Principle of a simple concentrator with a buffer near each terminal.

(as shown in Fig. 5.10). In this scheme the terminals are never locked, as the data is first transferred into the buffer and only from there to the concentrator. A request is sent to the concentrator whenever there is data in the buffer, and it is transferred at a speed which prevents any of the terminals from being locked. The advantage of this simple concentrator, relative to a multiplexor, is that the maximum line speed can be achieved and that it may flow directly into the computer without the aid of any intermediate equipment. In spite of the advantages that this type of concentrator can offer, its applications are limited, because special circuitry is required to be added to each terminal. This means in effect that the concentrator can function only with intelligent terminals. Finally, the buffer also adds the terminal address, to enable it to be identified at the other end.

The concentrators as described are simple but not completely satisfactory. A better solution is to transfer the buffering and all the control functions to the concentrator, thus releasing the terminals from any transmission function. Having the buffer at the concentrator end will completely separate the terminal from the line. All that the terminal does here is to transfer the character it transmits to the concentrator, where there is a buffer which is dedicated to each terminal (as shown in Fig. 5.11). The buffers, which are of the size of a character, are transferred to a second buffer where they form in a queue ready for transmission. This other buffer provides a waiting line for all the output messages. The advantage of this scheme is that it ensures that each terminal will always be able to unload its messages, because there will always be room for them to be held. The messages are transmitted to the line at the maximum speed, whenever the line becomes available, according to the principle of first come first served. Lastly, the logical circuitry in the concentrator must add to each message the specific address of each terminal.

The basic concept of the communication concentrator as described is in the buffering ability, which provides for the low-speed traffic arriving randomly to be held and then transmitted according to the line conditions. Thus the flow of data along the common line is smooth and at maximum speed. The principle of the concentrator which is based

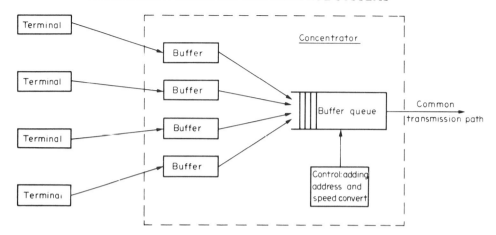

Fig. 5.11. Basic principle of a simple concentrator which utilizes a common queue buffer.

on the presence of a buffer is known as "hold and forward" or "store and forward".

Although the described concentrators present significant improvements in the efficient utilization of the transmission path, they still have some limitations. The main limitation is that an address must be added to each message which could be of the size of a character, and this reduces the efficiency by 50% if the message is also the size of a character. Another disadvantage is the size of the buffer, which must be economically of a feasible size and at the same time must be large enough so that no overflow conditions are produced.

5.6 ASYNCHRONOUS TIME-DIVISION MULTIPLEXOR

In the previous section the simple concentrator was discussed as one of the design approaches in increasing efficient line utilization. A different approach to the question of efficient line utilization is offered by the asynchronous time-division multiplexor, which is a hybrid between the time multiplexors and the concentrators.

The purpose of the asynchronous time-division multiplexors is to apply the time-division multiplex principle more efficiently. It combines the advantageous features of both the multiplexor and concentrator. The terminals are scanned as in multiplexing schemes, that is, consecutively, except that in synchronous multiplexing each terminal receives equal time slots, while in asynchronous multiplexing instantaneous time slots are used. The principle of the buffer storage operation in concentrators is applied also in the asynchronous time-division multiplexors, but whereas in the concentrator scheme the terminals request the line, this is not the case here.

The fundamentally inefficient operation of the synchronous time-division multiplexors can be seen in Fig. 5.12. There each terminal receives a fixed time slot on the transmission data sequence stream, and the terminals are scanned consecutively at a fixed rate. In other words, the line is connected to each terminal for the duration of the time slot and is transferred to the next terminal in turn at the end of the time slot. (As the scanning of the terminals is according to a fixed sequence, there is no need for an address to label each message transmitted.)

With synchronous time-division multiplexors the scanning of the terminals is

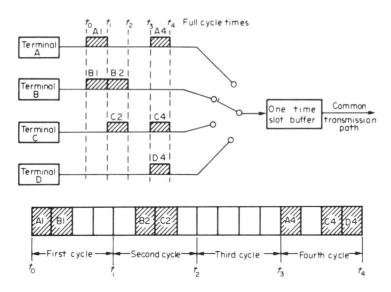

Fig. 5.12. The principle operation of synchronous time-division multiplex (STDM).

independent of whether or not there are messages available to be transmitted. This means that the multiplexor will remain connected to each terminal for the full period of each time slot, irrespective of whether it is required or not. This important point is demonstrated by the train of data emerging from the synchronous time-division multiplex that is shown in Fig. 5.12. It can be clearly seen that if there are no messages available for transmitting, then the specific time slot remains empty, that is, the time allocated to that specific terminal is not utilized but simply wasted. This creates an economic problem, since it has been proved by statistical measurement that only 5% of the time is used when communicating between terminal and computer, and not more than 35% of the time when communicating in the reverse direction.

To solve the random arrival of the messages, a more flexible scheme is required. This scheme would transmit the messages to the line in instantaneous sequence with no empty slot. It is this scheme which is known as asynchronous time-division multiplexing (ATDM).

It has been pointed out that in asynchronous time-division multiplexing the terminals are scanned in the same manner as in synchronous time-division multiplexing, namely, in a fixed sequence. The difference, however, is that if there are no messages to be transmitted from a specific terminal, the multiplexor will not halt but will immediately continue to the following adjacent terminal in the sequence. In this manner the multiplexor switches from terminal to terminal at a fast speed and will only halt at a specific terminal if there is a message to be transmitted. The time allocated to each terminal for transmission is fixed and is long enough for the terminal to unload the message, which could be of the duration size of a bit, a character or a block, depending on the multiplexor design. With this arrangement there is a great saving of time.

The principle of the asynchronous time-division multiplexor scheme is shown in Fig. 5.13, where the time saving can be seen from the data train sequence. The time saved by this scheme could be used for serving more terminals and is similar to that achieved by

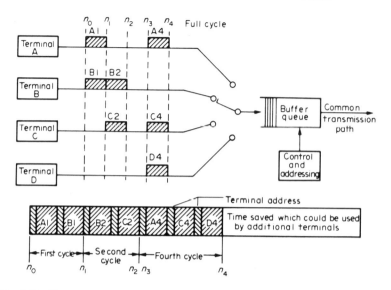

Fig. 5.13. The principle operation of asynchronous time-division multiplex (ATDM).

the concentrator discussed in the previous section. In both schemes, each is granted access to the common line whenever it has a message to transmit.

The four important features which characterize both the concentrator and the asynchronous time-division multiplexor are the following:

(a) The terminals are completely isolated from the common transmission line.

(b) Delays can occur between the time the terminals dispatch their messages and the time they are transmitted on the common line.

(c) An address must be added to each message to define the terminal the message originated from.

(d) A buffer store is required to hold the queue of messages which arrive at random times.

A labelling address must be added to the message as it is accepted from the terminals so that it can be identified by the computer. The message with the label is located in the b·ffer queue and then transmitted out at a constant synchronous stream. (This stream with the messages and address can be seen in Fig. 5.13). The buffer must be of reasonable size so as, on the one hand, to prevent long queue delays and, on the other hand, to prevent an overflow when the buffer is full.

The control unit supervises the operation of the sequence and provides means of encoding the message, i.e. adding the address and any other encoding data required. The multiplexing scheme is based on asynchronous arrivals of the messages and their synchronous transmission departures at a constant speed over the common path. The design of the buffer size is based on the statistical behaviour of the message arrivals analysed by the queuing model with finite waiting lines. This statistical analysis is based on the number of terminals, the volume of data from each terminal, the maximum allowable delay and the overflow probability. The design trade-offs are then calculated by the buffer size, the number of terminals scanned and the speed of both the input and output lines. The asynchronous time-division multiplex design, which is based on the statistical model, is known as Statistical Multiplexors.

5.7 STATISTICAL MULTIPLEXOR

In the previous section the asynchronous time-division multiplexor was discussed, showing how it improves the efficient utilization of the common transmission path. The scheme described was based on character interleaving time multiplexing.

The cost effectiveness of the communication network is related to the cost of the transmission lines and the cost of the modems and multiplexing equipment. Although the cost of the multiplexors could be high it is still small relative to the cost of the transmission lines. The whole design aspect is concentrated on the efficient utilizing of the lines by having the maximum volume of data flowing within the minimum period of time. Both the concentrator and the asynchronous time-division multiplexor produce this effect, but as has been shown, adding the address to each character message reduces the efficient utilizing of the line. Efficiency can be partially attained by increasing the message length, although it consequently increases the cost of the multiplexor. A compromise between these two extreme effects would be the production of an optimum multiplexor which yields the minimum operating cost yet satisfies the required performance. This could be achieved by means of what is known as a Statistical Multiplexor, which is broadly an asynchronous time division multiplexor with additional logic circuits.

As stated above, the design of the statistical multiplexor is based on a compromise of message length which will produce minimum queueing delay and also operate at minimum transmission cost. For this, three types of messages are considered (as shown in Fig. 5.14), viz.

(a) Constant short length messages.

(b) Random length messages.

(c) Segmented random length messages with fixed size blocks.

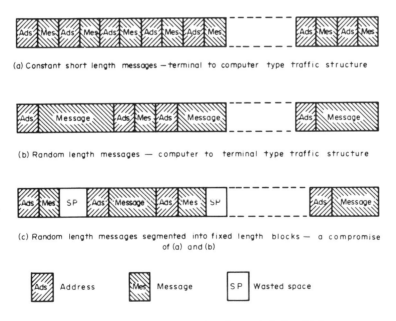

(a) Constant short length messages —terminal to computer type traffic structure

(b) Random length messages — computer to terminal type traffic structure

(c) Random length messages segmented into fixed length blocks — a compromise of (a) and (b)

Ads Address   Mes Message   SP Wasted space

Fig. 5.14. Message structure options for statistical multiplexors.

The design of the message length is also affected by the direction of the traffic flow. The terminal to computer traffic is generally of fixed size, usually of one or a few characters at a time. On the other hand, the traffic which flows from the computer to terminal is of larger varied size messages. The random arrival of the messages can be approximated by a Poisson distribution.

The constant short-length messages, shown in Fig. 5.14 (a), corresponds to the terminal-to-computer type of traffic, where a single character is transmitted for each time unit. This is a most efficient type of message for the multiplexor design and is best performed with a single common buffer where all the messages are waiting in a queue form. The data transmitted by this type of message is most wasteful owing to the relatively high percentage of the space taken by the address.

The random-length message, shown in Fig. 5.14 (b), corresponds to the computer-to-terminal type of traffic where a variable large number of characters string each message. The random nature of the variable length complicates the problem of storage allocation. Special circuitry will be required to identify the start and the end of a message, and the queue will take the form of a complex model each time it handles a different length message.

The problem of the variable message length could be met by segmenting the messages in fixed sized blocks, as shown in Fig. 5.14 (c). The fixed block size will simplify the buffer storage allocation, and thereby simplify the multiplexor design for computer-to-terminal traffic. Though the segmenting will cause some wasteful space with messages smaller than the block size, the design with segmentation of messages presents the most generally economical type of multiplexing.

The segmenting into fixed sized blocks could be performed in two schemes. The first builds the blocks immediately after the messages' arrival and transmits them regardless of whether they are full. This type of transmission could be effective for computer-to-terminal traffic, where the messages emerge in long bursts, but is uneconomical in the reverse direction, since the terminals produce the messages at a slow rate. In the second scheme, the blocks are transmitted only if they are full or only after a fixed time has elapsed, thus reducing the possibility of transmitting unfilled blocks. The second scheme will inevitably produce some delays in transmitting the terminal-originated traffic, but will produce no delays in the computer-originated traffic, since the blocks there are filled as soon as the messages emerge. This waiting period, despite delays, is the best compromise between performance and storage cost.

5.8 PROGRAMMABLE CONCENTRATORS

Statistical multiplexors are the most effective type of traffic concentrator, but once there is any overflow of traffic it may produce errors. Furthermore, each application may require different functions, such as editing facilities or error control. To incorporate all the possible required options in each multiplexor at a reasonable price is not always feasible. This would require a wide range of different types of statistical multiplexors to satisfy the market needs both technically and economically. The added flexibility of a programmable concentrator, as opposed to the fixed performance of the statistical multiplexor, enhances its usefulness by its ability to add processing. Wherever the traffic justifies using a concentrator, the designer should ascertain first whether a statistical multiplexor could solve all the requirements. Only if the available multiplexors cannot do so should he turn to a programmable concentrator.

The most appropriate description of the programmable concentrator is that it combines and utilizes both hardware and software techniques to perform communication concentrating functions. In other words, a programmable concentrator utilizes a computer with storage capabilities. The concentrator's function is to smooth the flow of data on the high-speed line by eliminating any transmission peaks which may occur when collecting the data from slow-speed terminals. The concentrator receives the data from a multiple of lines operating at different speeds and then transmits them at a constant speed after adding the addressing information.

The concentrator will accept data from a range of low- to high-speed terminals and concentrate their traffic on one higher-speed transmission path. The programmable concentrator can assemble complete messages and transmit them in blocks, thus relieving the load from the central computer. These points indicate the difference in performance of the concentrator and the statistical multiplexor. The statistical multiplexor usually handles traffic from terminals which transmit their data at a fixed speed (although at random times) and has limited editing abilities.

It is important to distinguish between the concentrator and the statistical multiplexor. The statistical multiplexor is a special hardware system designed to perform a specific job. Concentrators use a small general purpose computer which enables both hardware and software to be used to provide more facilities. Furthermore, the facilities in the concentrator can be changed by simply altering the programme, which is not possible in statistical multiplexors. In multiplexors the lines are scanned sequentially, while in concentrators the lines may be either scanned sequentially or operated by request.

As explained above, the programmable concentrator is usually a minicomputer which can present a large scale of service functions which are not provided by the schemes previously discussed. A concentrator of this kind can perform many functions, such as sorting, formating, error detection, correction and simple processing. In this respect a concentrator is similar in function to a small-scale message-switching system (a subject which will be discussed in Chapter 6).

It is also important to distinguish between the concentrator and message switching. The concentrator smooths the data flow from the terminals to the centre, and could be regarded as a traffic cop controlling the data transmission. With concentrators, the data flows in one or two directions only, that is, from terminal to centre or from centre to terminal. The object of a message-switching system is much wider than that of the concentrator, as it acts as a large-scale communication organizer. It not only directs messages to various locations according to the address of each message, but it can also reroute the messages along alternative paths and may, if required, even reconstruct the message itself. Whereas the concentrator serves as an intermediate stage between the terminal and the centre, the message-switching unit could be the centre itself. A further distinction is that the concentrator supervises the operation of the terminal associated with it while the message-switching unit only supervises the traffic of the message, and accordingly has negligible influence over the terminal operation.

The concentrator operates in a technique known as "hold (or store) and forward" which is distinct from systems which permanently store the message. This term may also be used for the statistical multiplexors.

Figure 5.15 shows the general configuration of a communication network which utilizes both concentrators and multiplexors in the same network.

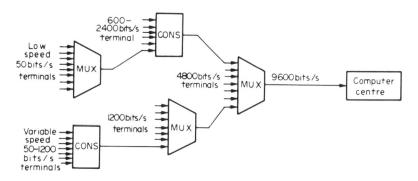

Fig. 5.15 Hypothetical example of a communication network hierarchy

5.9 CONCLUDING COMMENTS

The design range of multiplexors and concentrators is no longer confined to the traffic originating from a single terminal or even from a group of terminals, but must now be concerned with the volume of traffic in the whole network.

The geographical locations of each terminal in the network and the distances between them are the most important features in the design consideration. The first step is to specify the general network layout, itemizing the location of the local communication centres. The centres may utilize either multiplexors or concentrators, although at this design stage, no equipment should be defined. The volume of traffic from each centre may be calculated as the total of all the traffic originating from all the terminals associated with the centre.

If the number of terminals at each location is small, it may be questionable whether any common equipment is required. Separate lines direct to the nearest other centre should then be considered. If the distance to the other centre requires long lines, the cost effectiveness should be investigated by comparing separate lines with that of a single line with a simple FDM equipment. The break-even point depends on the cost of the lines, the distance and the cost of the common equipment.

For relatively small volumes of traffic, simple multiplexors should be considered. If the terminals are scattered over a wide area, FDM seems to be the solution. If the terminals are clustered together, the STDM seems to be most suitable.

With very large volumes of data traffic flowing in a complex communication network, then concentrating techniques should be pursued. If the data input from all the terminals is of a similar nature, then ATDM seems to fit the requirements, but if the input data is of a variable nature (i.e. different speeds and different type data) programmable concentrators may be required.

For the transmission over long distances with a large traffic volume, wide-band transmission channels are required. Here both FDM and ATDM should be investigated. The simplest and most commonly used system is that of FDM, but for better utilization of the bandwidth the ATDM is recommended.

5.10 REFERENCES

1. Martin, J., *Telecommunications and the Computer*, Prentice-Hall Inc., 1969, pp. 276-292.
2. Martin, J., *Teleprocessing Network Organization*, Prentice-Hall Inc., 1970, pp. 184-210.

3. Doll, D., "Multiplexing and concentrating", *Proc. IEEE*, Vol. 60, No. 11 (Nov. 1972), pp. 1313-1321.
4. Chu, W.W., "A study of asynchronous time division multiplexing for time shared computer system", *1969, Spring Joint Computer Conf. AFIPS Conf. Proc.*, Vol. 35, pp. 669-678.
5. Chu, W.W., "Demultiplexing consideration for statistical multiplexors", *IEEE Trans. Comm.*, Vol. 1 COM-20, No. 3 (June 1972), pp. 603-609.
6. Rudin, H., "Data transmission: a direction for future development", *IEEE Spectrum* (Feb. 1970), pp. 79-85.
7. Rudin, H., "Performance of simple multiplexor-concentrator for data communication", *IEEE Trans. Comm.*, Vol. COM-19, No. 2 (Apr. 1971), pp. 178-187.
8. Newport, C.B. and Ryzlak, J., "Communication processors", *Proc. IEEE*, Vol. 60, No. 11 (Nov. 1972), pp. 1321-1332.
9. Ball, C.J., "Communication and the minicomputer", *Computer*, Vol. 4, No. 5 (Sept./Oct. 1971), pp. 13-21.
10. Matteson, R.G., "Computer processing for data communication", *Computer*, Vol. 6, No. 2 (Feb. 1973), pp. 15-19.

CHAPTER 6

Switching Centres

6.1 INTRODUCTION

The discussion in the previous chapters was concerned only with the simplest communication network where the information flows from a terminal to a central location. In this condition all the communication network is operated on the principle of point to point with no provision for alternative routings. As the network grows, the efficiency is correspondingly reduced. Even the introduction of multiplexors and concentrators, which improves the efficiency of the common point-to-point path, does not ensure the overall efficiency of the complex communication network.

In sophisticated communication networks, the data flow is not only from the terminals to a centre; there is also considerable traffic between terminal and terminal. In large networks there could be a multiplicity of computers distributed over wide geographical locations, where there is a fair-sized traffic between the computers. This type of communication network configuration is common in Command and Control systems, for example, in law-enforcement systems and flight-control systems, where each airport may have its own separate computer network although also operating in association with all others for overall traffic control.

To achieve a comprehensive terminal-to-terminal communication in a simple network, each terminal must be connected directly to all others. That is, all the terminals have point-to-point lines between them, as shown in Fig. 6.1 (a), which illustrates the ability of all the terminals to communicate at the same time. This network (which is known as non-switched) is obviously impractical for large networks. Therefore, the only possible course is to bring all the terminal lines to a central location, as shown in Fig. 6.1 (b). The

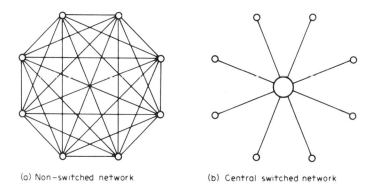

(a) Non-switched network   (b) Central switched network

Fig. 6.1. Principle of central switching.

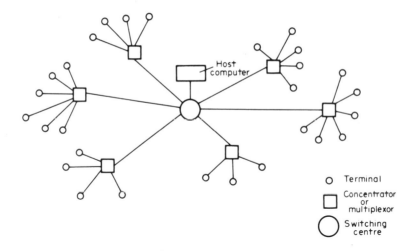

Fig. 6.2. Communication network with a single switching centre.

introduction of the centre switching exploits the whole network (and not just a single transmission path) and effects a saving in communication lines. The centre station is located at the hub of the communication network, linking all the remote stations in a configuration known as a star network. The remote stations may be either terminals, multiplexors or concentrators controlling a number of terminals, as shown in Fig. 6.2. The reduction in the long-distant lines has a very significant economical advantage, and furthermore, the centre can perform many functions beyond the capacity of point-to-point lines.

In early communication centre networks the switching was executed manually, as is still performed in a manual telephone exchange. Similar systems also exist for data transmission, and are referred to as "torn tape" switching centres. The writer is familiar with existing operating systems where a message is lifted off one teleprinter and transferred manually to another for the purpose of retransmission. These methods have proved to be most inefficient, and the current aim is to construct automated switching centres which can provide the following facilities:

(a) Elimination of human intervention in message handling.
(b) Increase of network efficiency.
(c) Economizing in the number of routes.
(d) Higher speed of transmission between centres.
(e) Maximum traffic throughout.
(f) High circuit utilization.
(g) Maximum common reliability.
(h) Simplicity of operation.
(i) Ease of alteration.

Centre switching is a step forward towards making the communication network more sophisticated. Clustering all the communication lines to a switching centre, besides being economical in regard to the network, is also advantageous by providing means of pooling the resources of the service organization. Clustering implies that a single maintenance staff can manage more equipment.

The centre's functions, as stated above, may be performed by a multiplexor or a concentrator; actually, however, they only perform traffic smoothing and do not fulfil any switching functions. The centre switching is usually produced by a stored programme computer which introduces a powerful tool with flexible programmes. Although concentrators use minicomputers, they are limited in the number of communication facilities they can present. Concentrators are intended to serve as a sort of traffic cop collecting data from various terminals and directing it to a common transmission path. An important feature of centre switching which is not found in concentrators is the facility with which any terminal can communicate with any other terminal. In addition, the system presents a multitude of links which enable alternative routings to be adopted when the main route is faulty.

As the traffic in the network increases, the efficient organization of the traffic and the network becomes a vitally important factor. The multiplexor and concentrator assist in the efficiency of a specific transmission line, while the central switching computer can deal with the traffic of the whole network. Furthermore, it can also act as a front end controller, preparing all the messages flowing in the system to be fed into the host computer. In this way the traffic does not load the host computer, which is left to perform its main task, i.e. processing. This aspect of centre switching is one of the major achievements, considering that all the communication control would take a lot of CPU time (as will be seen in Chapters 9 and 10). An additional feature is that the communication control is divorced from the information processing, both functions being performed by separate computers.

In large communication installations, such as Command and Control systems, there could be many such switching centres in the network (as shown in Fig. 6.3). The spread of computer resources over wide geographical locations requires local switching centres that can communicate with other remote centres. Each centre functions as a unique unit whose purpose is to switch and route messages from any of the terminals associated with it to any other associated terminal or centre in the network.

The introduction of means of switching and routing between centres promotes the efficient utilization of both the line and the equipment. It helps to locate the shortest path when the message has to flow through a number of centres, and to find alternative routing when one of the links is down. As the traffic and network increase, there is always a chance that a message will get lost *en route*. With the aid of the sophisticated

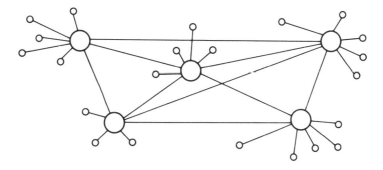

Fig. 6.3. Multi-switching centres–communication network.

centres, this problem could be reduced. Nevertheless, if the messages do get lost in transit, means are provided in centres both for recognizing the fact and for reproducing the lost messages. Where a number of switching centres exist in the same network, the network is open ended, with no single overall control. This type of network is important for some Command and Control systems as it is indefinitely expandable.

There are two basic concepts of switching and routing of the messages. The first is known as circuit switching. This sets a semi-permanent path within the centre, preserved for the duration of the transmission. The second is known as message switching. This routes the messages to their destination according to the address given at the head of the message.

6.2 MESSAGE SWITCHING VERSUS CIRCUIT SWITCHING

Switching of digital formed data can be performed either by circuit or message-switching systems. With today's techniques both effects are produced with the aid of general-purpose stored programme computers (as shown in Fig. 6.4). By programming, the computer causes the switching of the data between the input and the output lines, selecting the most appropriate available link. Despite the similarities of both centre switching systems, they use completely different techniques.

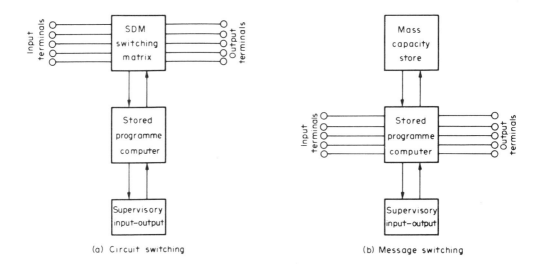

Fig. 6.4. The position of the computer in switching centres.

Circuit switching is most commonly used in telephone systems, where the best example is the automatic telephone exchange. In this centre switching technique, a circuit is switched in order to produce the required communication path. Once the path is available, the communication between the two connected terminals is direct. If the calling terminal cannot get through, because the called terminal is busy, the operator is obliged to close down and try again later. This unsatisfactory operation causes valuable time to be lost in making the connection. Furthermore, the operator has no means of knowing when the called terminal has become free. On the other hand, once the connection has been established, no message delays or queues occur. When the connection is no longer required, the corresponding link is freed and is made available for other required connections.

Circuit switching is extensively used for transmission of analogue type data but could also be used (under certain conditions) to transmit digital data. Originally, circuit switching was established with the aid of electro-mechanical type contacts, where the control is electro-mechanical or electronic. In the last few years, however, fully electronic digital techniques have been introduced, where the analogue data is transformed to PCM codes and the switching connection is performed with space division multiplex. So far, the fully electronic switching systems exist only in a few scattered working models, as the economics for preferring them to semi-electro-mechanical or semi-electronic systems on a large scale has not yet been fully established.

For data transmission systems, circuit switching is used when dialled connections are employed. For Command and Controlled systems, circuit switching has little application, except where the information transmitted is in the form of analogue data, such as in telemetry-type systems. Another possible application which could be useful is when connections are required for long duration so as to transmit digital measurement. These cases, however, may call for specially designed switching centres.

Circuit-switching techniques are fully dealt with in the literature and need not be discussed in detail in this book. Message switching, however, does call for attention, since it is the main interest of Command and Control systems, due to the fact that the messages transmitted between the terminals are relatively short and need to be transferred in "real-time".

In message switching all the data is received at a centre, where it is first stored, and only then can it be transmitted to the required destination. There is no direct contact between the sender and receiver as in circuit switching, and all the handling of the data is performed by the centre. The centre here deals with the actual message transmitted, or rather the heading of the message. It uses the message heading as a guide for transferring the message through the network to its destination, providing error control, and notifying the sender of the expeditious reception of the message. The operation of message switching for data transmission is a considerable improvement over the circuit switching, as the feasibility of achieving a connection is no longer a matter of chance. The messages that are transmitted to the centre are always accepted there, irrespective of whether there is an output route available to retransmit them to their destination. Message switching requires no physical path to be set between the sender and the receiver; and consequently, no special equipment resources are needed for each connection. It is clear, then, that the best equipment efficient utilization for data transmission is obtained in message switching.

Multiplexors (FDM and STDM) could be regarded as primitive circuit switching, and concentrators as primitive message switching. Circuit switching is usually applied for applications where long connection times are required, while message switching is intended for short contact periods.

Some of the differences between circuit switching and message switching are presented in Table 6.1.

Another point which should be considered when comparing circuit switching with

TABLE 6.1

Facilities	Circuit switching	Message switching
Means of communication	Direct between terminals	Indirect with aid of the centre
Form of data	Analogue or digital	Digital only
Connection time	Relatively long	No time lags
Delay in connection	Delayed till a path can be located	No delays at terminal end
End to end delays	None once the connection is made	Depends on the queue of messages
Destination terminals busy	Requires closing down and a second try	Of no interest to the calling terminal
Utilization of the lines	Independent of the traffic	Smooth traffic flow
Efficient usage of equipment	Inadequate efficiency	Maximum efficiency
Sharing equipment resources	Dedicated to each connection	Common to all connections
Data handling	None whatsoever	Storage, routing and error correction
Data storage	Not available	Available for back checking of messages
Application	Long transactions	Short enquiry – response transactions
Connection information	Separate from message	Message heading

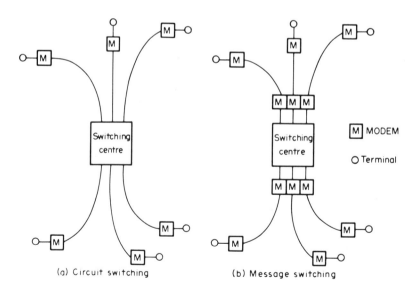

Fig. 6.5. The requirement of modems in a switched network.

message switching for data-transmission applications, is the number of modems which are required for each system (as shown in Fig. 6.5). Since circuit switching can handle both analogue and digital data, there is no need to convert the transmitted data when it passes the connection matrix. The modulated media of the data are already in a form which could be handled by the system. In message-switching systems, however, the data must be only in digital form; hence, all the data flowing in the lines must be demodulated before it can be handled by the system. This will result in two modems being required for connection between two terminals in a circuit-switching system, and at least four modems in a message-switching system.

As already mentioned computers can be applied for both circuit and message switching, with advantages dependent on the practical applications. In the case of pulse code modulation transmission, circuit switching seems to be the most profitable, since here the connections are made over long periods and no advantages would be gained by commoning the path (except those gained by the PCM principle). In this application the advantages gained with using circuit switching are that no delays occur once the connection is made, and the connection can be held for long periods. The initial delay at the beginning of each communication transaction is of no relevant importance, since it is negligible relative to the actual contact period. In computer to computer communication applications, message switching is the most appropriate, because the messages are extremely short and require brief contact time. There is, however, a serious obstacle to be overcome: the initial delay is still relatively high compared with the short contact period. For real-time applications, where the messages arrive at random times but must be transmitted instantaneously, the initial delays in making the contact could be a matter of vital importance affecting the successful operation of the whole system.

6.3 MESSAGE SWITCHING

It must be clearly noted that the message-switching centre is designed to control the transmission of digital information from one remote location to another. Message switching is essentially a combination of equipment and techniques for obtaining high speed, accuracy and time saving. In the centre, the message heading is checked, the address is identified and then the message is routed over the first available path to the appropriate destination. The system is intended to permit a smooth traffic flow of messages, without any human intervention. It is a dynamic technique which is constantly being improved to meet new requirements.

The most important feature of message switching is that the terminals are not directly connected to each other over a fixed path but only indirectly, through the centre with which all the terminals are linked. The centre is alone responsible for the forwarding of the messages and must always be available to accept the messages from the terminals. In other words, the terminals can transmit their data to the centre at the maximum speed irrespective of whether there is an available output route from it. It is obvious that the centre must have a high-speed computer which is capable of obtaining the data from many input lines simultaneously. This means, then, that the system uses real-time programmes which provide flexible operations. Message switching must operate on a 24 hours a day basis in an environment where only the highest standards of reliability are acceptable.

Message switches are used in three distinct fields of applications. In the first, the centre acts as an administrative control of traffic between various computer installations. In the second, the centre controls the traffic between a central data-processing system and a wide geographically dispersed terminal organization. In this latter application, the message-switching system acts as a front-end processor where on one side it controls the traffic from all the incoming lines and on the other prepares the data to be fed into the central computer. In the third field, message-switching acts as a central communication coordinator to provide message communication within a single organization. In all three fields, the functions presented by the system are similar, although the first application field is the most sophisticated. The two other fields of application use a simple form of message switching, employing a single central switching computer installation similar to the star configuration which has already been discussed. (The computer to computer communication will be further discussed in Chapter 10).

The messages transmitted must indicate in the heading the addresses of both the destination and the sender. This will enable the centre to identify and define the required route. As the message need not be of a fixed length, the heading must include an indication of the message length.

The messages are transmitted to the switching centres, where each message is recorded on a suitable medium. The routing information which is contained in the message heading is analysed. If the outgoing route is free, the message is transmitted, but if busy, it is held until the route becomes free. Before this operation, it is first determined whether the address is directed to a valid destination. If it is not, the sender must be informed that the address is incorrect. (This operation is most important when the messages flow through a number of switching centres and an incorrect address can cause a message to get lost in the system.) If the message cannot be transmitted immediately, the message is kept in backing store until a free route can be located. The system keeps on trying to find the required route and the alternative routes until an available route is located. The sender is not informed of any delays in dispatching the messages, unless the waiting period is in excess of a specified time.

Each message is given a priority value within a given scale. This value will be recognized by the centre, which will then proceed accordingly. High-priority messages will always be transmitted before those lower in the scale. Very high priorities, if so required, could stop all operations so as to free a transmission link. The system as described might cause low-priority messages to be left in the queue indefinitely, but this state is avoided by suitable means, whereby the priority value of messages is altered, after a given period, to enable them to be advanced within the switching centre.

As the messages reach the centre they receive a unique number. This facilitates the checking of the message to ensure that it is not lost *en route*. The messages originating from each terminal work on a running number scheme which allows easy checking, manual and automatic, to see whether a message from a particular terminal has been lost. Within the centre, the numbering scheme is much more complicated and special log tables must be formed to pursue the messages flowing between a number of centres.

No processing is involved in message switching, and the computer is used only as an organization tool. The computer must operate in a real-time environment, accepting data from various lines simultaneously. That is, the centre must function so that it can accept the messages at all times from all terminals and in unlimited volumes. The input lines are actually scanned sequentially at a very high speed, which gives the effect of receiving the messages from all the input lines simultaneously. In systems where a host computer is associated with the message-switching centre, the communication computer prepares the data in a form which could be handled by the host computer. In this configuration, the communication between the communication and host computers is effected by using an interrupt scheme, after which the data is transferred at a very high speed.

All the data arriving on the lines is first stored in a buffer which is associated with each line. The buffer size is usually one character long, as larger sizes would be too expensive for a very large number of lines. The buffers are scanned and the data is transferred to a mass capacity store. The scanning operation is similar to a multiplexor operation, where the ATDM scheme would seem to be the most profitable. The messages are kept in the storage medium until a line becomes free. This characteristic, which was also found in the concentrator, is known as the store and forward technique.

An important feature of message switching using store and forward techniques is that the messages could be held for long periods after the message content has been dispatched. That is, the messages may be retained for indefinite periods in the large capacity auxiliary store. If a query subsequently arises concerning the contents or destination of any messages, the messages may then be easily retrieved. The messages are stored according to the message reference number or/and the date. Normally the messages are requested to be retained for 24 hours, but there are systems which may require far longer periods. This addition of storage capacity could be quite a burden on the cost of the system.

The star network with a single switching centre may be quite adequate for many applications, but as the system expands, it demands the introduction of long parallel transmission lines, and this will result in the escalation of the communication cost. Furthermore, any failure in the switching centre will disrupt the whole communication network. It will then be advisable to install a distributed switching network to effect a decentralization of the resources in such a way that any failure does not disrupt the whole network but only isolated local terminals. In this configuration (as was shown in Fig. 6.3) the transmission of the messages between the terminals associated with one centre is the same as before. The connection between terminals linked to different centres could also operate in the same manner, provided that the address of all the terminals can be recognized by tables held in each centre. The programme would determine from the address in the heading of the message whether that message is local or one which must be transmitted to other centres. Although the trunk lines between the centres are usually of higher speed than those to the terminals, the message-switching centre handles them in the same way as it does the local lines.

The main function of message switching is to transmit the message to its destination according to the address at the heading of the message. The address may be one of three kinds: simple, multiple or group. A multiple address is used when the message has a multiple destination with each individual address located in the heading. A group address is a single address which has a multiple destination located according to a predetermined table.

In some systems, such as telemetry, the messages may arrive with no address at all. In these applications, the centre may add the address according to predetermined tables applied to the source of the messages. In more sophisticated systems, the centre may add extra destination addresses in accordance with the specific contents of the message, apart from that given in the heading by the originator. This function is important for management applications, for there a number of personnel must be kept informed as to what messages are flowing in the system. In this application the processor scans the message for key words according to a predetermined table and then adds the new address information to the message heading without notifying the sender.

6.4 FACILITIES PROVIDED BY MESSAGE SWITCHING

The basic purpose of message switching being to communicate messages from one point to another, all the other facilities which can be presented by the system are clearly derivative. The number of these facilities is unlimited, since they can be introduced into the system by appropriate programming of the computer. Each application may require different facilities, with the choice depending only on the memory capacity and the operation cycle time. Some of the accepted facilities are stated below:

(a) Provision for receiving messages at all times, from remote terminals or from other centres, all possibly arriving simultaneously.

(b) Routing of messages to a number of terminals in parallel at a maximum speed.

(c) Automatic speed conversion, receiving messages from various lines operating with multiple speeds.

(d) Automatic routing of messages, according to the address, to single and multiple destinations. That is, parallel distribution of the messages to all the terminals, to some of the terminals, or to a single terminal.

(e) Provision for dealing with messages of variable length.

(f) Provision for handling messages arriving with or without the address information and from equipment which sends synchronous or asynchronous type messages.

(g) Storage of the message until the destination line is free.

(h) Provision for storage of all the messages for indefinite periods of time, for further study if required, and to ensure that nothing is lost in the system.

(i) Arrangement of the outgoing messages in a queue according to a priority scale, with provision to change the priority if the message is held too long. In complex systems, the highest priority messages could be given preemption rights with the interruption of messages already in transmission.

(j) Automatic self-check of the system network for reliability of both the lines and equipment. When a fault is located, automatic transfer of equipment and transfer of the traffic to alternative lines. The supervisor is informed of any breakdown, while the users are informed if one of the terminals is faulty and cannot receive any messages.

(k) Automatic error detection and correction.

(l) Validity test of message heading to determine whether the destination is valid or if any other part of the message heading is valid.

(m) Maximum message throughput, i.e. automatic selection of messages for transmission, as soon as heading has been identified and analysed, even before the full message has been received.

(n) Editing the messages, with facilities to add, delete or to change format. In complex systems, the editing could also apply to the actual message text.

(o) A statistical analysis of the traffic, to locate bottlenecks and to present suggestions for future development of the network. The system provides various statistical reports based upon information from collected logs, input/output traffic, terminal requests and alarm conditions. The system can also provide full history of all the logs and produce traffic graphs according to queries.

(p) Adding numerical information to check the reliable transfer of the messages. This numerical information could be used both for statistical purposes and as a tag for retrieval of the messages.

(q) Automatic code and format conversion. This involves the translation from one procedure to another, since there are a number of different communication procedures that could operate in the same system and this may require the use of different speed, character set and control procedures. In fact, different procedures are generally used when communicating with terminals, other centres and the host computer.

(r) Hard copy facilities, for whatever is required.

6.5 MESSAGE SWITCHING – CENTRE STRUCTURE

The structure of the centre of a message system generally consists of a duplex processing network. In this arrangement both halves of the system are in operation, but only one processor is active, the other being at standby. The conventional duplex system requires complete duplication of all the equipment vital to the processing task so that if any unit fails in half the system, then the complete half system is instantly failed and the other half of the system takes over. This is the most simple duplex arrangement, and is known as system standby, as shown in Fig. 6.6 (a). A much better arrangement is to interconnect modular units of one-half system with those of the other half, as shown in Fig. 6.6 (b). This arrangement, known as modular standby, will enable the system to continue full operation even with a number of faulty conditions. (A more detailed discussion of the modular configuration is given in Chapter 14). The full modulator approach is expensive and so is not used in most of the message switching systems currently available.

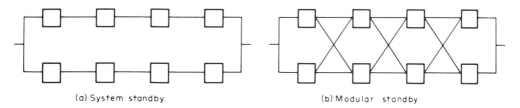

(a) System standby   (b) Modular standby

Fig. 6.6. Standby facilities.

A typical simple duplex processing message-switching system is shown in Fig. 6.7. Both halves of a dual system accept incoming messages, process them and place them in the output queues. Nevertheless, only the active processor can output the messages to the lines, while the other processor output is inhibited. Each processor has its own memory and peripherals attached to it; accordingly both systems may operate separately in a real-time environment, where both processors operate simultaneously, building up identical information patterns in the memory.

The automatic change-over unit is a separate equipment which checks the current active processor operation and switches over to the standby processor when a failure is recorded. In effect this unit continuously monitors all equipment both in the active and in the standby processor. With a failure in either system, the supervisor is informed, but

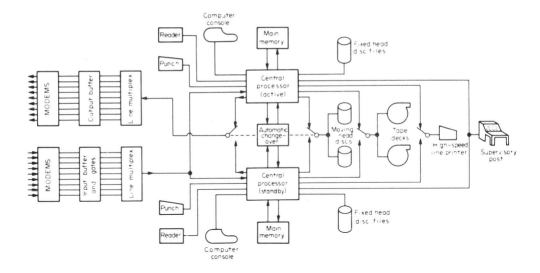

Fig. 6.7. Centre block diagram of a message-switching system.

only when a failure is recorded in the active processor is a switch-over performed. The automatic change-over unit is an equipment which continuously receives OK pulses from various parts of the processors indicating the successful operation of the processors. Should a failure develop, one of the OK pulses will be inhibited and a failure will be recorded. If this failure occurs on the active processor, the automatic change-over unit will direct the standby processor to become active by switching over all the auxiliary equipment.

The automatic change-over unit must also keep both processors in synchronization, on a block basis, to assure that no information is lost on switchover. As the standby processor always obtains a duplicate of the input data on all lines and builds up an identical data base, the traffic is not disrupted when a change over takes place. When the standby processor becomes active, it uses the information in its own data base to reconstruct the recent history of processing, and can then resume normal operation.

The data is received from and transmitted to the lines via line control modules which handle it on a character-by-character basis. The line control consists of simple character buffers. These lines are scanned in the same manner as in asynchronous time division multiplexors.

The memory organization is built around a three-level hierarchy: main, backing and auxiliary memories. The main memory is provided by a core or by bipolar random access arrays, which are intended to hold the currently active application programmes. The backing mass capacity memory is provided by a drum or by a fixed-head disc file, which is intended to hold all the current data files being processed. As the messages arrive, they are assembled into blocks in the main memory unit and then transferred into the backing memory. In this arrangement, the complete message, heading and text, is held in the backing memory. As soon as the message heading is received, it is returned to the main memory to be analysed and verified. As a result of this analysis, the procured routing information and the messages are then placed in queues on the relevant outgoing lines. The messages are not moved from the backing memory when they are placed in the queue; instead a message pointer is placed on a suitable log table. Once the lines become free, the messages are lifted off the backing memory by the processor, a block at a time, and transferred back to the main memory. From there they are transmitted to the line, via the line-control module, a character at a time. After the messages have been successfully transmitted to their appropriate destinations, there is no need to continue to hold them in the backing memory. That being the case, the copy of the dispatched messages can be transferred to the auxiliary memory for long-term storage. The auxiliary memory may be provided by tape decks or/and by moving-head disc files. The auxiliary memory plays no active part in the successive transfer of the messages and, in consequence, is of no operational importance. For this reason, there is no need to duplicate the auxiliary memory, as was the case with the main and backing memories, and a single unit can serve both processors.

The instructions are fed into each processor separately, to enable them to operate independently. This also allows the service of one processor, with the other processor otherwise engaged. This requires that all the peripheral equipment should be duplicated and assembled separately on each processor. The relevant peripherals that may be attached to each processor are paper tape and/or card reader and punch which are used for on-site programme feeding, testing and maintenance. Each processor must also have its own console with a keyboard typewriter, to communicate between the operator and the specific processor. On the other hand, the high-speed line printer, used for the high volume statistical reports, may be shared between both systems, as it plays no part in the message handling.

The supervisory console must be common for the full duplex processing systems, as the same operator is responsible for the whole system operation. That is, the system must be regarded as a single structure, it being of no importance for the message transfer which processor is in operation.

In any automatic communication system the attention of the supervisors and that of the terminal operators must be drawn to unusual operational effects which are not anticipated. The supervisor provides interrupt handling, error recovery procedures, and supervises all the control of the centre operations. The computer must be able to recognize those circumstances which require attention, and request the intervention of the supervisor. The supervisory console unit must be provided with facilities for instructing the computer to take whatever actions are required to deal with the problems involved. The supervisor must also be able to request information from the computer on the operational status of the system. Some of the supervisory functions are listed below:

(a) Facilities overriding all the automatic functions.
(b) Inserting extra data into the system.
(c) Retrieval of all data from the system.
(d) Inserting checking messages.
(e) Change of routine priorities.
(f) Change of routings.
(g) Change and add programmes.
(h) Handling all the alarms.

The alarms in a message-switching system must be extended beyond those found in a time-shared real-time system. The proper functioning of the communication network depends entirely on the quick attention of the supervisor to any alarm conditions. Some of these alarms are itemized below:

(a) When the terminals called do not react.
(b) Lines are disconnected.
(c) A message is delayed in the system above the allocated time.
(d) An ultra high priority message is in the system.
(e) Security violations.
(f) Various hardware and software faults.

6.6 THE FUNCTIONAL CHARACTERISTICS

As already explained, the centre must accept messages from distant terminals in real-time and store them in the system. On the receipt of a message, the heading is analysed. The complete heading information being only needed for the processor operations, it need not be transmitted in full context to the final destination, but it is needed when the message is transmitted to another processing centre. The heading must contain all the relevant information required for the analysis. The items are generally grouped together and transmitted before the message text. The heading information depends on the application requirements, some of the items of which are listed below:

(a) Start of message (SOM) character, which should essentially precede each message, since all the messages arrive at random times. This character, when recognized, will initiate a sub-routine in the processor to accept a new message.

(b) Message serial number, which is essential as a reference to locate the message held in file. This serial number can provide means for detecting failures in the successful dispatch of the messages. If the serial number is in sequence the loss of the message is detected as soon as the following message arrives. In other numerical schemes, a summary of all the messages received and transmitted from each terminal is compared with the log kept in the centre of all the traffic.

(c) The date and time of sending the message are important items which assist in locating old messages in the files. They are of interest not only to the terminals but also to the centre, which requires them for an historical analysis of the traffic. In multi-centre systems, the date/time can help in defining the duration of the messages delayed in the system, and thus enables the priority of over-delayed messages to be changed.

(d) Message-length character, which specifies the number of characters of the accompanying text. This item of information is important, as the messages are not all of the same length. Once the full text of the message has been received, the system can verify with this character whether the full message has been received. With the aid of the character the system may prepare suitable storage capacity to hold the message. This information is only required for the centre processing operation and need not be transmitted to the final destination terminal.

(e) Priority character, which places the importance of the message in a scale of prominence. (The priority handling has already been discussed.)

(f) Destination address, which must contain enough information so that the centre can determine the correct communication path for retransmission of the message to its destination. The route is determined by reference to stored routing tables, after checking the address to find whether it is valid. This is particularly important in multi-centre systems, where an invalid address would result in the message getting lost in the system. In the case where the message is addressed to multiple destinations, the number of destinations must be limited so as not to overload the system. The validity check of the address must also be performed, to see if the number of addresses is not excessive.

(g) Sender's address, which is important when there is an error in the message heading necessitating the return of the message to the sender. In some applications the sender is also informed when the message has been successfully dispatched. This information is important also for statistical analysis, to record all the messages originating from one installation.

(h) The sender's address is sometimes processed by a character code indicating the operator's specific name. This code must be verified against a clearance list stored in the computer, to ensure that no unauthorized messages can be inserted into the system. If an unauthorized person operates any terminal, the supervisor is immediately notified of any violation of security.

(i) A message text is defined as all the characters following the heading information. Since the heading length may be of variable length, the text must be preceded by a start of text (SOT) character, with an end of message (EOM) character following the text, and recognized by the system as initating processor functions, and for checking any errors in message length.

The computer programmes which handle the messages and process the heading can be divided into seven categories (as shown in Fig. 6.8), viz.

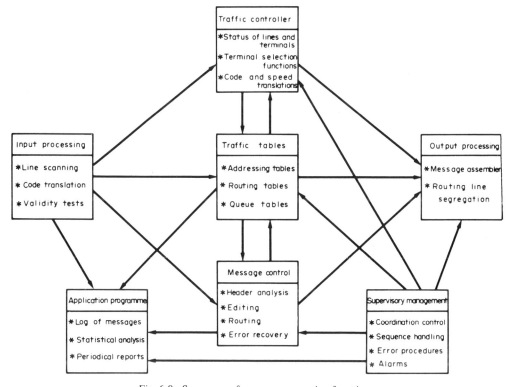

Fig. 6.8. Summary of message-processing functions.

(a) Input processing programme

Its routine functions are to scan the lines and organize the received messages for processing. It determines the type of traffic, mode of operation and, where required, initiates codes translation and formating routines, i.e. converts input, external, terminal, independent codes into a standard computer code. It performs validity tests, security tests and error detection. These programmes must analyse the heading and initiate other programmes as soon as the heading is available, even before the full message has been received.

(b) Output processing programme

Its routine functions are to check all the outgoing lines, to allow a smooth flow of data. It assembles the messages for transmission on the lines in the correct code format and speed. The output programme must also have functions which ensure that the messages are accepted within a given time duration.

(c) Traffic-controller programme

This deals with all the traffic problems which are not concerned with the messages themselves. It performs terminal selection functions, device-control functions and buffer managements. It is responsible for speed conversions and synchronization requirements. It checks the status of lines and terminals, and logs the excessive noise on the lines. It notifies the operators when lines go down and updates the tables of the status of all routes available.

(d) Message-control programme

This is the main programme of the system, responsible for the handling of messages in the processor. It analyses the message header, defining the routes to the appropriate destinations. It creates the message queues according to the priority of the message for each route. A special programme edits the message, changing the heading where required, including change in priority. If a message is delivered with an error, it could initiate the block recovery for retransmission.

(e) Supervisory management programme

This is the main programme for coordinating all the programmes and arranging the sequence of all the operation controls and peripheral device operations. It also performs error procedures associated with computer failures. These programmes must also handle all the alarm conditions and allow the supervisor to override the automatic functions.

(f) Traffic tables

These are the heart of the system operation, as they consist of all the permanent data required when handling the message switching and routing. Some of these tables are listed below:

    (i)   Address table of each terminal and the routing information.

    (ii)   Line table, with indicators of the terminal or centres assembled on it.

    (iii)   Users' table, with identifying codes for security test.

The tables also consist of logs of non-permanent information which is essential for proper operation. They contain the queue of each output line where a special table is required for each one. Other tables in the system contain the dynamic locations of the message blocks in the backing memory.

(g) Application processing programmes

These programmes contain all the functions not required to be operated in real-time. They keep logs of all the messages which pass through the system, and enable retrievals of old messages on request. They perform statistical analysis of the traffic and provide periodic reports on the operation. They handle all enquiries regarding message and traffic conditions.

6.7 REFERENCES

1. Martin, J., *Teleprocessing Network Organisation*, Prentice Hall Inc., 1970, pp. 233-245.
2. Hamsher, D.H. (Editor in Chief), *Communication System Engineering Handbook*, Chapter 7 by P. Schneider, "Message switching engineering", pp. 7-33 to 7-42, McGraw Hill Book Co., 1967.
3. Richards, R.K., *Electronic Digital Systems*, John Wiley & Son Inc., 1966, pp. 481-518.
4. Richards, R.K., *Digital Design*, John Wiley & Son Inc., 1972, pp. 563-571.
5. Newport, C.B. and Ryziak, J., "Communication processors", *Proc. IEEE*, Vol. 60, No. 11 (Nov. 1972), pp. 1321-1332.
6. Mills, D.L., "Communication software", *Proc. IEEE*, Vol. 60, No. 11 (Nov. 1972), pp. 1333-1341.
7. Mathison, S.L. and Walker, P.M., "Regulatory and economic issues in computer communication", *Proc. IEEE*, Vol. 60, No. 11 (Nov. 1972), pp. 1254-1272.
8. Doll, D.R., "Multiplexing and concentration", *Proc. IEEE*, Vol. 60, No. 11 (Nov. 1972), pp. 1313-1320.
9. Kahn, R.E., "Resource-sharing computer communication networks", *Proc. IEEE*, Vol. 60, No. 11 (Nov. 1972), pp. 1397-1407:
10. Dell, F.R.E., "Features of a proposed synchronous data network", *IEEE Trans. Comm.*, Vol. COM-20, No. 3 (June 1972), pp. 499-503.
11. Matteson, R.G., "Computer processing for data communication", *Computer*, Vol. 6, No. 2 (Feb. 1973), pp. 15-19.
12. Rudin, H., "Data transmission: a direction for future development", *IEEE Spectrum* (Feb. 1970), pp. 79-85.
13. Decker, H., "Communication processing for large data networks", *Data Processing Magazine* (Nov. 1970).
14. Doll, D., "Planning effective data communication system", *Data Processing Magazine* (Nov. 1970).
15. Ball, C.J., "Communication and the minicomputer", *Computer*, Vol. 4, No. 5 (Sept./Oct. 1971), pp. 13-21.
16. "The 6400 ADX system series for automatic message and data switching", Standard Telephone and Cables Ltd. (UK), ITT, Aug. 1972.
17. "IBM system 5910 message switching system", GA 19-5000-1, 1971.

CHAPTER 7

Communication Network Hierarchy, Architecture and Control

7.1 INTRODUCTION

In this chapter the discussion will be centred on the aim of reducing the overall cost of the complete network and not, as in the previous chapters, of only a single transmission path. As the best communication system design for maximum utilization of the network is the linkage of line control with the centre equipment techniques, the chapter will discuss means of controlling the traffic on the lines and taking advantage of the facilities presented by the intelligence available in the centres. It will also discuss the network topology, showing the design tools needed in defining the actual placement of the centres and means of communicating between them.

A communication network system must accommodate many kinds of data concentrating techniques with different levels of sophistication. The growth of a data-communication system calls for a hierarchy of communication techniques in order to smooth the traffic flow in the network. These techniques are intended to achieve the maximum traffic flow of messages in minimum time. The hierarchic structure of the network techniques is illustrated in Fig. 7.1, and shows four levels.

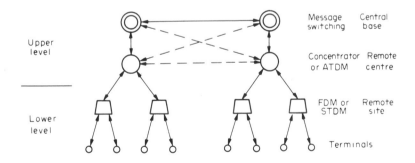

Fig. 7.1. Hierarchy of communication techniques.

Before continuing the discussion, it is essential to define the general term used for a network junction point. Each point of the network convex where the network interfaces with the outer world or with other network points is known as a node. The node controls the outgoing traffic on the lines connected to it. In this reference a node may be a terminal, multiplexor, concentrator or message-switching unit, although the terminal is seldom used in this role. The hierarchic structure is defined by the node characteristics and not by its physical placement in the network. Thus the connection between the nodes need not necessarily be according to their position in the hierarchic organization. For example, a terminal may be connected directly to the central message-switching computer, or a frequency division multiplexor may be connected with both the concentrator and the message-switching computer (see also Fig. 10.2 on page 204).

The hierarchic organization of the network techniques can be divided into two distinct groups, according to the characteristics of the nodes. A crude division of the hierarchic structure classifies the lower level nodes as 'dumb' and the upper level nodes as 'intelligent'. In the lower level the nodes act only as message carriers, while in the upper level the nodes operate as message manipulators. A further distinction is that the message handling in the lower level nodes is performed with hardware techniques, whereas in the upper level nodes the handling is done with software techniques.

The lower-level nodes in the hierarchic structure use frequency division multiplexors and synchronous time division multiplexors. In these nodes the transmission path is divided into a fixed number of channels, where each channel is dedicated to each input terminal. These channels are always available to their associated terminals, thereby eliminating delays in the message handling. On the other hand, these nodes are most inefficient in line utilization, and furthermore, they do not perform any communication control, such as speed or code conversion.

The upper-level nodes in the hierarchic structure use asynchronous time division multiplexors, concentrators or message-switching systems. These nodes all operate in a store and forward mode and share the transmission line on a requirement basis. The upper-level nodes use intelligent equipment, i.e. computers, which introduce further communication facilities and provide better and more efficient line utilization. These nodes can furnish routing functions, speed conversion and format conversion. Set against the advantages of the upper-level nodes are the unsatisfactory features that they can produce delays in the message handling and that they call for expensive equipment.

7.2 NETWORK ARCHITECTURE

In most Command and Control system organizations the terminals are scattered over large geographical areas. In the previous chapters the discussion was concerned with the design of the centre nodes, where the data is collected from remote sites and transmitted

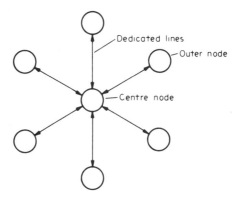

Fig. 7.2. Principle of the star network configuration.

to its destination along a single common transmission path. That is to say, the network organization, so far discussed, considered a communication scheme known as 'multipoint', where each of the connecting lines between any two nodes utilizes a separate line. In this multipoint scheme the network configuration operates on point to point communication techniques between the nodes. The general multipoint network configuration is based on a star network (as shown in Fig. 7.2) where the centre of the star is a switching communication node and the outer nodes are remote sites. Here the centre and the outer nodes of the star network may be either of the same or different level in the hierarchic structure. While the centre of the star may be any type of node except terminals, no restrictions are imposed on the outer nodes. For instance, the centre node may be a concentrator and the outer nodes terminals. Likewise, the centre of the star may be a message-switching unit, while the outer nodes may be concentrators. In the latter example, each outer node of the star may be a centre of a separate star. In all these network configurations, each node is connected to the centre, where two outer nodes can communicate only via the centre node. The connection between the outer nodes and the star centre can be regarded as a separate point-to-point operation.

The multipoint star network is by no means the most economical network configuration, since it calls for a dedicated line between each outer node and the centre node. Where the outer nodes of the star are concentrators or multiplexors, the dedicated lines to the main centre are a common transmission path for all remote outer nodes of a secondary star, whose centre is an outer node of the main star (as shown in Fig. 7.3). By

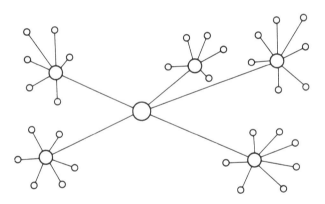

Fig. 7.3. Double star configuration where each outer node of the primary star is a centre of a secondary star.

commoning the traffic on a single line, the traffic flow on the line is smoothed out; however, if there is not enough traffic originating from any remote outer node, the line will not be fully utilized. In consequence, the multipoint network suffers from underexploited lines. A much better network arrangement is to have a number of outer nodes on the same line in a configuration known as a multidrop network scheme. A multidrop network utilizes a single transmission line from which all the outer nodes drop (as shown in Fig. 7.4). The transmission line runs from the centre node through all the remote nodes to the most distant node.

The question arises when to use a star network and when a multidrop network. The answer is to design a communication network which consists of a combination of both

Fig. 7.4. A typical multidrop transmission line.

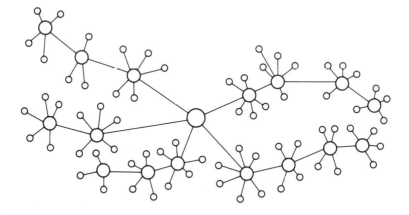

Fig. 7.5. A typical communication network configuration which utilizes both star and multidrop techniques.

star and multidrop techniques (as shown in Fig. 7.5). The selection of the configuration technique to be used depends on the distance between the nodes and the traffic volume which originates from each node. As the distance between the centre node and the remote nodes increases, the design tendency should be towards the application of a multidrop configuration. However, if the traffic volume originating from a remote node is large enough to dominate a line, then this node will require a separate transmission line to the centre node with no other nodes dropping from it.

The two network configurations discussed, i.e. star and multidrop, are widely used in communication networks for lower hierarchy level nodes. With the further increases of the node sophistication, network topologies are used to include the refining facilities the nodes can provide. For example, the network architecture using intelligent nodes can provide alternative routing between the nodes, which is not possible with the lowest level nodes.

The communication organization can provide a wide choice of network configuration, but it is essential to distinguish whether the remote nodes are in the lower or upper level of hierarchy. Each of the topology configurations presents efficiency advantages but also poses problems in line control and error recoveries.

Where the outer nodes are 'dumb' equipment, the network branch configuration must be centralized. A centre in this respect may be either a multiplexor, concentrator or message switching which can serve a number of incoming lines. The outer nodes in these branch configurations may be terminal, multiplexor or concentrators. As already explained, the topology configuration for these lower hierarchic level nodes can be only one of two geometric patterns, as shown in Fig. 7.6, viz.

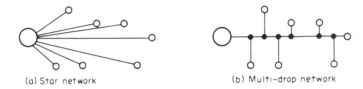

Fig. 7.6. Alternative network configuration for connecting remote nodes to a centre (where the remote nodes are not message-switching centres).

(a) Star network, where each remote node is directly connected to the centre, with a separate transmission line.

(b) Multidrop network, where all the remote nodes are connected to the same transmission line.

In the upper hierarchic level the nodes provide more facilities, and each node may be part of another network. The nodes in this aspect are all computer type and so present different types of network topology. Multidrop is not used in these configurations, as it is not practical to have a number of remote computers on the same transmission line. The alternative network geometries for the higher hierarchy level nodes are shown in Fig. 7.7.

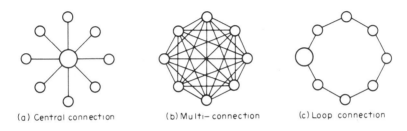

Fig. 7.7. Alternative network configuration for computer to computer communication.

They comprise:

(a) Star network, which is a highly centralized scheme where all the data flows to the centre node. Although the outer nodes are computers, the connection between them is only via the central computer. This configuration is unreliable, as the failure of the centre node will cause a complete breakdown. Though this type of failure is also possible in the lower-level nodes, the damage there is not so critical as it is for computer to computer communication.

(b) Multiconnection network, which is a highly distributed scheme with no specific centre. Each node (which must be a message-switching centre) can be connected to each other node and so increase system reliability by providing alternative routing. The network configuration may require all the nodes to be fully or only partially connected. A variation of the multiconnection is the chain network, where all the nodes are connected in series. The messages there are transferred from node to node till they reach their destination.

(c) Loop network, where all the nodes are connected in rings. The transmission line threads through all the nodes in series, starting from the centre node and returning to the centre node in a closed loop. This scheme is a variant of both the centralized and distribution schemes. The traffic flow is controlled centrally (although the message handling is distributed) enabling direct communication between the outer nodes. This topology network will be discussed separately in the following chapter.

7.3 MULTIDROP NETWORK CONFIGURATION

In the previous section it was established that where the traffic from a remote node is relatively small, a separate line to the centre is not economically warranted. In these cases a multidrop configuration could present the best cost effective utilization. In this network configuration all the outer remote nodes are connected to the same communication line, where the network acts like a telephone party line. Each node is assigned an identification code, i.e. an address, which is unique to each node. When the centre sends out a message headed by the individual distinctive code, only the node the message is intended for will accept it, while all the other nodes are inhibited. The remote nodes must include hardware for recognition of the individual address transmitted on the line.

The multidrop topology links all the nodes in the shortest available path between them, and this is exceedingly more economical than the star topology, as seen in Fig. 7.8 (a). Each line drop is not restricted to a single node, but the line may branch out with a number of nodes dropping from each branch. This configuration gives an even more economical configuration, as seen in Fig. 7.8 (b).

Traffic in a multidrop configuration flows only between the central node and the

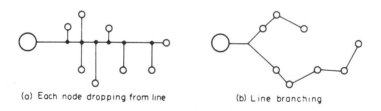

(a) Each node dropping from line   (b) Line branching

Fig. 7.8. Multidrop network with Branching.

remote nodes, with no direct communication between individual remote nodes. Only one outer node can communicate at any given time, while all the other nodes must wait till the line becomes available. The allocation of the line facilities can only be provided by the centre node, which acts as a traffic cop controlling the use of the line by the various remote nodes. The centre node generally polls the individual remote nodes and asks them whether they have any messages to transmit. The remote nodes may transmit only when the centre nodes give them the "green light". As only one node may transmit or receive at any given time, it calls for complicated line-control discipline methods. (These will be discussed separately.)

The main operational constraint of multidrop networks is that delays can build up in the remote nodes while waiting for the line to become available. Delays may be caused when a single remote node dominates the line by continuously transmitting data. This problem can easily be avoided by the line discipline procedures, which allow each remote

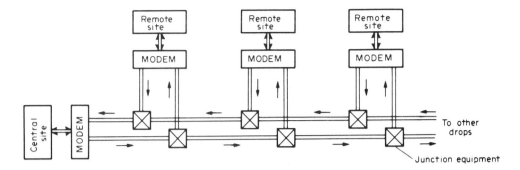

Fig. 7.9. A four-wire multidrop network configuration.

node to occupy the line for a period long enough to transmit only a single block. Delays can also occur when all the remote nodes have messages to transmit and each node must consequently wait its turn. In consequence, during the busy periods, if too many nodes drop from the line, long queues can build up in the nodes until the line becomes free. If there are considerable overflow conditions in the remote nodes, separate lines must then be considered.

Each node, remote or centre, is connected to the line via a modem (as shown in Fig. 7.9). The lines which are generally used for multidrop are four wires which may tolerate full duplex. Although full-duplex lines are provided, most multidrop applications use the line in a half-duplex configuration procedure, with one line for data and the other for control purposes (as was shown in Fig. 4.6(b)). Nevertheless, there are many multidrop configurations which use two wire lines in a half-duplex mode.

An important operational constraint of multidrop is in the hardware connections of the nodes with the line. Each junction connection will change the line characteristics, thus reducing the transmission speed and increasing the turn-around time. The junction adds many communication problems, such as increasing noise level, linearity, envelope delays, attenuation and frequency response. Special interfacing equipment must be mounted at the junction in order to equalize the line characteristics of the three line outlets. To allow proper compatibilities for data transmission techniques, the junction equipment usually consists of amplifiers and hybrid circuits. The introduction of equalizers does not solve all the line problems, as they can be adjusted only to optimize performance over individual line characteristics. For this reason it is advisable to have the equalizers near the modem and to ensure that they are adjusted automatically. These equalizers should be internal to the modems, but if this is not already the case they must be added externally.

7.4 MULTIDROP NETWORK DESIGN CONSIDERATIONS

The design of multidrop lines must consider not only the most economical type of network but also the load on the line and the reliability of the network. It must be kept in mind that multidrop lines are inevitably unreliable, as any disconnection on the line will cause the detachment of all the remote nodes situated beyond the uncoupled point.

The cheapest multidrop network and most economical configuration is one which gives the lowest total line distance, as shown in Fig. 7.10 (a), but it also presents the most

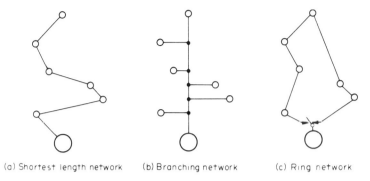

(a) Shortest length network   (b) Branching network   (c) Ring network

Fig. 7.10. Alternative multidrop network configuration which differ in cost and reliability.

unreliable network. A more reliable network (although also far from perfect) is shown in Fig. 7.10 (b), where each node is separately connected to the line. The disconnection of any individual branch will affect only one node and will not disrupt the full network operation. However, this network configuration requires longer total line length, and so is more expensive. Furthermore, its reliability still lacks the perfection required, since any disconnection of the common part of the line will disrupt the line operation. A much better network configuration is one which connects all the nodes in a ring, as shown in Fig. 7.10 (c). This alternative network configuration is more reliable than both (a) and (b) configurations. It is cheaper than the (b) configuration but more expensive than the (a) configuration. (The ring network operation is completely dissimilar to the loop network operation and the reader must be careful not to confuse them.) The centre node output may be connected to the transmission line either in a clockwise direction or in a counter-clockwise direction but never in both directions simultaneously. Furthermore, the ring must operate in an open-loop mode. When the transmission multidrop line is disconnected at any point, then, and only then, may the line be regarded as two separate multidrop branches which are fed from the duplicated output of the centre node. The centre must then feed the two halves of the line in parallel. In this case there is no need to change the line-control procedures, as the same number of nodes are operated by the duplicated output point. With a special simple diagnostic programme the exact location of the link disconnection can be easily located.

The design procedure of a multidrop network is based on the following aspects:

(a) Volume of traffic originating from each remote node.
(b) The transmission lines available.
(c) The transmission speed to be used on the lines.
(d) The physical location of the remote nodes.
(e) The network reliability.

The aim of the multidrop configuration is to reduce the total cost of the lines connecting widely scattered remote nodes. Nevertheless, this arrangement must consider the locations of the nodes and not assemble together nodes which are in opposite directions relative to the centre node.

The total number of nodes that can be assembled on each line will depend on the sum of the traffic from the nodes and the line speed. It must be remembered that about 40% of the maximum line capacity is used up by control procedures, leaving only 60% of the line volume to be taken into account for the actual information transfer. The volume from each node and the speed are calculated in bits per second. In these circumstances the average number of remote nodes that can be connected to the multidrop line is as follows:

Line speed × 60%/Average node output volume = Average number of nodes.

This equation would be simple if all the remote nodes were TTY terminals where the average output volumes are all approximately equal, but if the remote nodes are concentrators with different loads, then each node will feed the line with a different volume.

If the volume output from all the nodes in the line causes overflow, the line must be split into two (as shown in Fig. 7.11). This new configuration is a combination of a star and multidrop geometrics. The two separate lines should then be designed to give balanced shared loading on both lines.

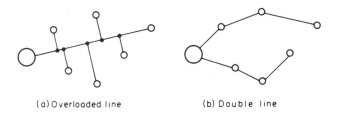

(a) Overloaded line   (b) Double line

Fig. 7.11. Balancing loads over two multidrop lines.

For all complicated network designs, a combination of a star and multidrop configuration must be considered. The design examination must aim at the optimum network structure, to present the lowest cost.

The design procedure should start with a star network where each remote node is directly connected to the centre node, as shown in Fig. 7.12 (a). The direct links to the centre are removed, one at a time, and a new link is established via another node by creating a multidrop line, as shown in Fig. 7.12 (b). Each time a link is added to the newly created multidrop line, the total volume of traffic on the line must be checked to ensure that no overflow will occur. Furthermore, a balanced system must be projected, with the same number of nodes on each multidrop line, as shown by the final circuit shown in Fig. 7.12 (c). The complexity of the network configuration options is wide, and it is suggested that the choice be made with the aid of a computer programme, so that all

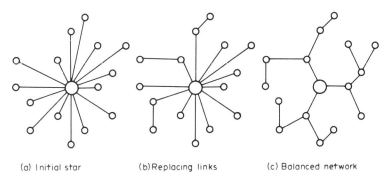

(a) Initial star   (b) Replacing links   (c) Balanced network

Fig. 7.12. Building a simple tree structure communication network.

the network configurations can be considered. With all the requirements met, the resulting network should present the cheapest network.

It should be realized that it is not only the maximum volume traffic and the link length which affect the choice of the network. An important design feature to be considered is the delays between the time the message arrives in the remote node and the time it can be transmitted to the centre node. These delays are created by the amount of messages each remote node has to transmit, and also by the number of other nodes in the line. Even if all the other remote nodes have no messages to transmit, a delay will be established at each node. This occurs because each remote node is scanned in turn by an inquiry requesting whether it has a message to transmit. Before the scan operation can be transferred to the following node, the remote node in question must send a reply announcing whether it has or has not a message to transmit. Each time this is effected, the line must wait for the turn-around time. Furthermore, each extra node on the multidrop line will add to the response time of the network.

In the circumstances the design direction must be modified to consider the following constraints:

(a) Balancing the total volume of the traffic in each multidrop line.
(b) Balancing the total length of each multidrop line.
(c) Balancing the number of nodes on each multidrop line.
(d) Balancing the total delays in each multidrop line.
(e) Minimizing the network cost.

The constraints of minimizing both the total line length (i.e. network cost) and the delay in dispatching the messages do not always tally in presenting a fixed network-design procedure. To give the designer scope tolerance, one of the two constraints must be defined by its maximum upper limit. The other constraint can be variable, checking each time that the upper limit of the first constraint is not passed. The design may be based on an allowed maximum average delay whilst aiming at the minimum network cost. On the other hand, the design may be based on the maximum accepted network cost while aiming at achieving minimum delays.

With the basic tree available, the design procedure can continue by considering the two constraints. Remote nodes may be transferred or exchanged between different multidrop lines. The network that was shown in Fig. 7.12 (c) can be improved by transferring nodes from branch to branch, as shown in Fig. 7.13. It can be seen that any tree can be obtained by a sequence of such transformations. After each transformation the total cost and the delay must be checked, for the node alliance transfer is feasible only if there is a

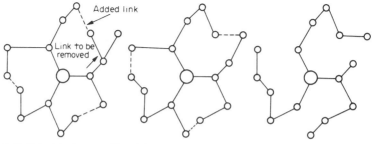

(a) Initial state to be modified   (b) Intermediate state   (c) Final state

Fig. 7.13. Building new multidrop trees by exchanging nodes between adjacent multidrop lines.

reduction in either of them. The design algorithm should start with considering the node farthest from the centre and building a temporary link with the closest node to it situated on a different multidrop line. Either of the two nodes can then be transferred from line to line. If the transfer is feasible, a new link is added to one multidrop line and the original link to the other line is removed. The second design step is to consider the next farthest mode on the original multidrop line. When all the nodes on one line have been considered, then the next line should be examined, but, this time, excluding all the nodes on the first line.

In the same manner, complete multidrop lines could be deleted, thus optimizing the network configuration. If the load on each line is not near its limit, extra nodes may be added to it. Each time a node is added, it must be checked as to whether the load, delay or cost make the transfer feasible. In this fashion, new trees are formed with the number of multidrop lines reduced, as shown by the example in Fig. 7.14.

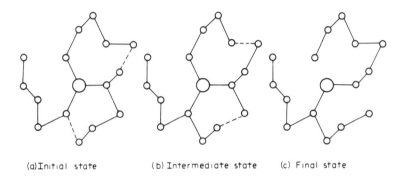

(a)Initial state   (b) Intermediate state   (c) Final state

Fig. 7.14. Minimizing multidrop network by transferring nodes from line to line and deleting obsolete links.

It was explained earlier in this section that multidrop lines are inevitably unreliable, and it was suggested that the network be arranged in rings. The rings must always be open loops, since each node must be defined in only one particular multidrop line. Furthermore, there must be no connections between different multidrop lines. The simplest ring network, shown in Fig. 7.15 (a) and based on that given in Fig. 7.13 (c), is a distributed network where some nodes are split between two rings. When a failure occurs on any single link, the operation will not be disrupted, as all the nodes could still be

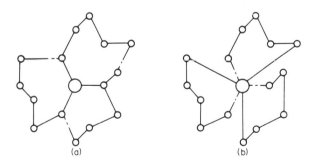

(a)   (b)

Fig. 7.15. Alternative multidrop ring configurations to prevent disturbances due to a link failure.

reached from the centre. However, the junction nodes will require local switching so that they can transfer their allegiance from one multidrop to another. Furthermore, the centre must then change its programme and file tables, because the address of the transferred nodes has to be modified, owing to the fact that they will be located on a different multidrop line. A much better arrangement, although more expensive, is to have each ring individually connected to the centre, as shown in Fig. 7.15 (b). In this configuration, all the switching will be centralized and there will be no requirement for any address changes during failures.

7.5 LINE-CONTROL PROCEDURES

In all communication network configurations a dialogue must take place between each remote node and the centre node. The question arises as to what line-control procedure should be used when connecting the central node with any of the remote nodes. It is obvious that strict rules must be enforced in the network to allow the message to flow at a maximum speed with no messages lost when the line is busy. The designer is confronted with a range of line-control procedures which are used at different levels of the communication hierarchy. The more sophisticated the nodes in the network are, the more complicated will be the control procedures, including restrictions on the traffic flow.

Most of the control procedures are based on scanning the remote nodes in sequence and allocating the line to a single node at a given time. The selected node will then have full and exclusive use of the line for a restricted period of time.

The various line control procedures available to the designer are listed below.

(a) Synchronous time sharing, where each node is presented with a fixed time sequence in which it can transmit its messages. This time is allocated independently of whether there are messages to be transmitted or not.

(b) Asynchronous time sharing, where each node is scanned sequentially to see whether it has a message to transmit. Only if the scanned remote node has a message to dispatch will the node be allocated a time period to transmit its message. The scanning of the lines and the allocation of the time period are provided by hardware equipment.

(c) Polling time sharing, where each of the nodes is asked to see whether it has a message to transmit. When a remote node has a message to dispatch it will be allowed to do so by a particular instruction sent to it from the centre node. The polling of the nodes and all the line-control procedures are provided by software programmes.

(d) Random time sharing, where the line is fed by the central node with a sequence of time slots. Each remote node in the network having an equal chance to capture any time slot, the seizure of the time slots is regarded as random. In this line-control procedure the remote nodes can communicate between themselves with no intervention required from the centre node. This technique is used in the loop network communication which will be discussed separately in the succeeding chapter.

(e) Contention control procedure, in which each remote node can initiate a request for line control. If the line is busy, the remote node joins a request queue to wait until the line becomes free. This is a distributed responsibility scheme, while all the other schemes are centralized.

Polling and contention can only be effective if the centre node is an intelligent organ. Furthermore, these schemes will also require special hardware equipment at each remote node. Synchronous and asynchronous time sharing require hardware equipment only at the centre, but will not require any special equipment at the remote nodes.

In polling and contention, the centre allocates the line resources. In polling, the centre initiates all the transmission procedures, with no initiation power presented to the remote nodes. In contention, it is the remote nodes which initiate the transmission of the messages, although here too, the centre node gives the authority permitting the remote node to transmit its message. In both schemes only one node can transmit at a given time, while all the other nodes are frozen.

In the contention scheme, the remote nodes send the centre node a request to dispatch. This request is like a flag which is raised and then recognized by the centre node. The request is placed in a virtual queue until the line becomes free. The allocation of the common line is presented to the remote node according to a prearranged priority scale at the centre node and the sequence of its appearance. The requests from nodes on the same level will be served according to the rule of first-come-first-served, but those of higher priority will still be served first. When a node is allowed to transmit its message, it will be restricted in time so as not to dominate the network. Contention is widely used in computer systems where the common line is the input bus and each request will cause an interrupt. Full duplex lines are essential for contention operations, where one line is intended for transmitting the request flag and the other line for half-duplex data. This arrangement is essential, as the remote nodes must be given a chance to raise a flag while the line is engaged.

Polling and asynchronous time sharing are very similar in their effects, that is, in the scanning of the remote nodes and in the allocation of the time slots only to the nodes which have messages to dispatch. In both schemes the remote nodes have no means for initiating any line-procedure operation, which is essentially a central-control operation. If any node in the network has a message to dispatch, it has no means of drawing the centre's attention. The remote node must wait 'patiently' until the centre asks the node whether it has a message to dispatch; and only then can it get rid of its 'burden'. In both schemes only one node can obtain line control at any given time.

Despite the similarity between polling and asynchronous time sharing, these two schemes are completely different. Asynchronous time sharing scans the nodes using hardware techniques, while polling uses pure software techniques. To initiate polling, the centre node must be an intelligent medium, while ATS requires sophisticated but not essentially intelligent equipment. ATS is intended for star network applications, while polling is for multidrop network applications. In other words, with ATS an individual line is required between the centre node and each of the remote nodes, whereas with polling all the lines to the remote nodes must meet at a single point at the centre node.

7.6 POLLING AND SELECTING

Polling and selecting are the most commonly used line-control procedures where the central node, which dominates the line operation, is a computer. The polling of the remote nodes may be effected by a remote concentrator, a message-switching centre, communication controller or a central processor. The existence of a remote concentrator or a communication controller can ease the task of the central processor by freeing it to perform more useful tasks.

Polling is used where individual point to point control in a multi-point network is impractical or uneconomical. A polling network allows several remote nodes to share an input communication channel to the central node. Although initially polling is intended for multidrop configurations, polling could also be used in a star configuration, provided all the incoming lines meet at a single point near the input to the central node. There are many node polling control procedures, such as IBM's BSC for synchronous communication, or Bell's B.A.1 for asynchronous communication.

Polling may be regarded as another method of multiplexing where the central computer scans the remote node sequentially, requesting data. If the elected remote node has no messages to dispatch, the succeeding node will be asked. If one of the checked remote nodes has a message to dispatch, the complete line control will be transferred to this node for a fixed period of time. The scanning of the nodes need not be according to a fixed sequence, as in time-division multiplexing, but in any sequence required. Furthermore, this scheme allows certain remote nodes to be scanned in each cycle more frequently than others and in any appointed sequence, with provision for easily changing the sequence. The sequence is according to a polling list which is controlled by a software programme stored in the memory. The polling sequence takes into consideration the priority of the nodes and the amount of traffic which originates from each node.

The centre node acts as the master station of the network, controlling all the traffic flowing to and from it by sending out suitable operation commands. The remote nodes act as slave stations, as they cannot initiate any line operation but must respond to all the centre's commands. (The relation between the centre node and the remote nodes is shown in Fig. 7.16.) There are two types of commands. The first applies where the centre

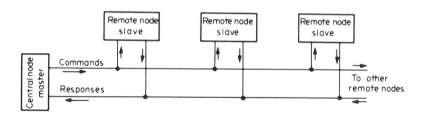

Fig. 7.16. Master–slave relationship between centre and remote nodes in a polling network.

node asks the remote nodes whether they have a message to transmit, and the second where the centre node itself has a message to transmit to a particular remote node. With these two types of commands, the centre node can control the direction of the transmission on the line. The definition of the two command operations is summarized as follows:

(a) Polling, in which the centre node scans the remote nodes to see if they have a message to transmit. In this case, the command is an invitation to the remote nodes to send a message.

(b) Selecting, in which the centre node elects a particular remote node for the transmission of a message to it. In this case, the command is an instruction to the remote node to receive a message.

Each remote node is allocated a unique code which contains sufficient information to make automatic response when it appears on the line. This unique code is known as the remote node address. The address information flowing in the line will reach all the nodes in the common network, but only a single node may respond to its unique address. All the nodes will tend to decode the address, but only the node addressed will be able to connect itself to the line. Accordingly, this operation is essentially polling and not scanning.

A special synchronous code is sent at the head of each command message which will alert the nodes to check the proceeding address characters. If the message is not intended for a particular node, it will recognize this by the address, and will relax its attention till an end-of-message code is transmitted in the line which tells it to wait for the next synchronous code.

Each remote node, on the receipt of a centre enquiry intended for it, must acknowledge. If it has a message to dispatch, it can then transmit it. If it has no message to transmit, it must also send an acknowledgement indicating that it has no message to transmit. When the centre receives a message from a remote node, it must acknowledge that the message has been received correctly or, if an error was recorded, to send a request for retransmission of the message. The operation indicates the main constraints of the polling and selecting, which are the following: (a) a large portion of the transmission time is wasted on the questions and answers, and (b) delays are encountered between each question and answer for the turn-around time period.

The following procedure schemes are introduced here to demonstrate, with the aid of an example, how line control is performed. The scheme should be regarded as an initiation version and not as an authoritative definition of the actual line-control procedures. The main control characters used for this example are:

SYN   Synchronous character, which precedes the transmission for the purpose of obtaining or maintaining synchronization.

EOT   End of transmission character, which marks the final end of a transmission sequence which may contain any number of messages. In multidrop lines, this character is also used as a signal to all the remote nodes to inform them that the previous communication to a particular node has been completed and that a new polling command may follow.

ENQ   Enquiry character, used to solicit a response from a remote node.

ACK   Acknowledgement character, used to affirm the positive reception of a message with no errors detected.

NAK   Negative acknowledgement character, used to request retransmission of the last block when a transmission error was recorded.

STX   Start of text character.

ETX   End of text character.

SOH   Start of heading character.

BCC   The total number of characters included in the actual message, that is, the message heading and text but excluding any synchronization characters.

The operation sequence of polling and selecting can be explained with the aid of the example shown in Fig. 7.17. Before each transmission message a SYN code is sent from the originating node to the receiving node for the purpose of obtaining synchronization. The EOT code is used to call the attention of all the nodes to be ready for an addressed command. With the transmission of an EOT followed by a particular address, a virtual

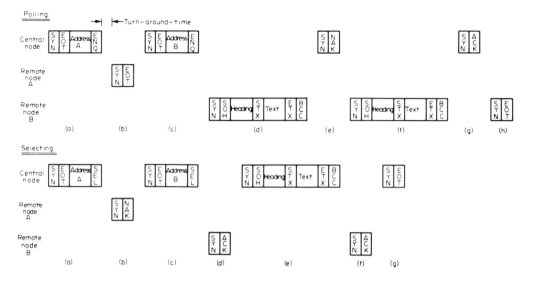

Fig. 7.17. Simplified version of a polling and selecting sequence of events.

point-to-point connection is established between the central node and one of the remote nodes. All the other nodes will be inhibited for the duration of the transmission connection, which is maintained until the next EOT code is transmitted on the line. The polling sequence of the example is as follows:

(a) The central node sends a message headed by a SYN code which synchronizes all the remote nodes. This is followed by an EOT code which sets all the remote nodes on the line in preparation to decode the proceeding address. The address characters which follow select only one remote node while inhibiting all the other nodes on the line. Following the address, a command code is transmitted, which in the case of polling is an ENQ checking whether the remote node has a message to transmit.

(b) The remote node which was addressed must reply to the centre's command. This reply is again headed by a SYN code, but it will only synchronize the central node, since all the other nodes are inhibited until the next EOT code. If the addressed node has no message to transmit, it will return a negative reply. There is no need to send a NAK code, as it could adequately send an EOT code which means that the transmission has been completed with no messages to transmit.

(c) The central nodes send another poll enquiry message to a second remote node at another address.

(d) The new remote node has a message to dispatch, which it transmits immediately after a SYN code. To save time there is no need to send a positive reply (ACK). The message transmitted consists of first the heading and then the text. The message ends with a BCC code which specifies the number of characters in the message.

(e) The central node checks whether there are any errors in the received message. If any errors are detected, the central node will then transmit a negative acknowledgement (NAK code) which would be recognized by the remote node as a command for retransmission of the message.

(f) The remote node retransmits the original message.

(g) The central node rechecks the retransmitted message and if no errors are recorded it transmits a positive acknowledgement (ACK code).

(h) The remote node on the reception of the ACK code can clear its buffer, since the message has been successfully transferred to the central node and so there is no need to hold it for error enquiries. The remote node having completed its tasks will transmit an EOT signal on the line which will tell all other nodes that the transmission has been completed and that they should be on the alert for a new poll on select command.

The selection operation is identical with the polling operation except that the command is now SEL, which informs a particular remote node to expect a message. The remote node must acknowledge such a command, either positively or negatively. A negative acknowledgement means that the addressed remote node is busy with other tasks. A positive acknowledgement means the node is ready to receive the message.

Instead of a particular command for polling or selecting, this instruction could be contained in the address itself.

7.7 DESIGN CONCEPTS FOR OPTIMIZING COMMUNICATION NETWORKS

In this section the design of the location of the centre node and the requirement of the links between remote nodes will be debated. In the discussion so far, and in particular with regard to the approach described in Chapter 5, the network topology has defined a link or a centre even when there is not enough traffic to warrant it. This design approach is not satisfactory, as in most cases it does not present the most efficient network topology. The question that must be answered is when and where a centre node is required and when should a common link be used.

As has already been argued, the design constraints of a communication network are based on the following considerations:

(a) The average volume of traffic originating from each remote node.

(b) The transmission lines available between the remote nodes (i.e. speed).

(c) The cost of the transmission lines.

(d) Type of equipment available at the centre and remote nodes.

With the system growth, centre nodes and multidrop lines will be used to common the traffic originating from a number of remote nodes. The design constraint must be extended to consider the following aspects:

(e) Balancing the traffic volume in the network.

(f) Compensating for peak traffic.

(g) Balancing time delays in the network.

(h) Balancing network cost.

(i) Balancing network reliability.

In any widely spread communication network there are numerous choices for arranging links between all the nodes. A computer programme is usually essential in considering all the possible design topologies. Even then it is doubtful whether the optimum network would be reached. The approach suggested here is to reduce the design variables to a minimum by considering the basic network structure.

In any communication network there is no general need for direct communication links between each pair of nodes, although traffic may flow between all of them. The

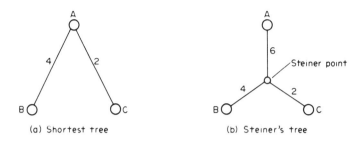

Fig. 7.18. Minimum network configurations (figures indicate number of channels).

design approach must consider traffic flowing through a number of remote centres before reaching its destination. Where no centre exists, a new centre may have to be introduced.

A communication network consisting of a number of links is known as a communication tree. The shortest tree structure is based on direct links between the nodes. A simple example of the shortest tree is shown in Fig. 7.18 (a) where nodes B and C are connected to a central node A. All the traffic flowing in the network is between A and B, and A and C, with no traffic between B and C. The shortest tree network is not always the most economical. This can easily be seen, as node A is at a further distance from nodes B and C. Nevertheless, even if equal distances exist between the three nodes, the shortest tree does not present the minimum configuration. By commoning the traffic to and from B and C nodes on a single link, the total sum of all the links in the new network will be less than that of the total sum of the links in the shortest tree. The new tree structure, shown in Fig. 7.18 (b), is known as a Steiner tree. This network creates an extra node, known as a Steiner point, which corresponds to the location of a multiplexor or concentrator. The same design approach could be applied where there are more nodes in the tree. A second example is shown in Fig. 7.19, which presents a tree network with four nodes. In general, if there are n nodes in a network, then there could be up to $n-2$ Steiner points in the new network configuration.

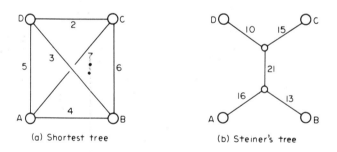

Fig. 7.19. Optimizing a four-node tree configuration.

Optimizing the geometric design of a tree network depends on the following factors:
(a)   The number of nodes in the network.
(b)   The distances between the nodes.
(c)   The number of parallel channels on each link.
(d)   The cost of the links between the nodes.
The number of channels N in any single link is determined by the volume of traffic

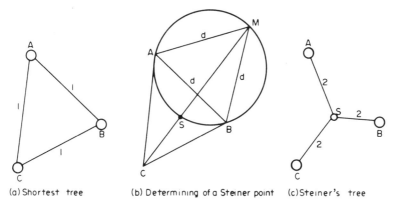

(a) Shortest tree   (b) Determining of a Steiner point   (c) Steiner's tree

Fig. 7.20. Constructing a minimum Steiner tree from the shortest tree.

flow in the link between its two nodes. The function $f(N)$ represents the cost per mile for installing the N channels. If the cost of the link is independent of the number of channels, the function $f(N)$ remains constant. In the example shown in Figs. 7.18 and 19, the number at the side of each link represents the number of parallel channels in the link. The cost $f(N)$ is not a direct function of the number of channels, as it also includes preliminary costs for surveying, digging and installing. For this discussion these items will not be taken into consideration.

The construction of the Steiner point may be performed with the aid of a ruler and compass. Consider a simple case of three nodes ABC in a tree network, with equal volume of traffic flowing between the nodes, as shown in Fig. 7.20. A line is drawn between any two nodes, e.g. $|AB| = d$, and a triangle is constructed with equal sides, e.g. ABM. The base of this triangle $|AB|$ must lie between the top point of the triangle M and the third node C. The triangle ABM is circumscribed by a circle and then a line is drawn between points M and C. The crossing-point of the line $|MC|$ and the circle represents the Steiner point S of the minimum tree network.

It is not in all the tree networks that a Steiner point can be constructed. If the segment $|CM|$ does not cross the circle or does not cross segment $|AB|$ then there is no minimum length solution. These two design constraints are shown in Fig. 7.21.

With tree networks with the number of nodes $n > 3$, the minimum length tree is constructed stage by stage. A circle is circumscribed round a triangle constructed by any

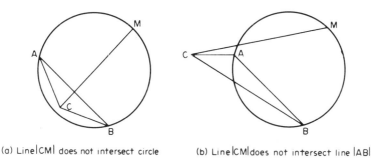

(a) Line $|CM|$ does not intersect circle   (b) Line $|CM|$ does not intersect line $|AB|$

Fig. 7.21. Examples where a Steiner point does not present the minimum length solution.

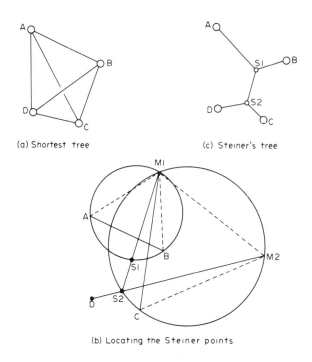

(a) Shortest tree

(c) Steiner's tree

(b) Locating the Steiner points

Fig. 7.22. Constructing Steiner points for a network with $n = 4$.

two nodes, e.g. $|AB|$. This circle will also consist of points M_1 and S_1. Then a second circle is circumscribed round segment CM_1 which will also consist of points M_2 and S_2. If the network consists of $n = 4$, as shown in Fig. 7.22, a line can be drawn between M_2 and D, and S_2 is located, where the circle crosses it. Another line can be drawn between M_1 and S_2 and this in turn locates S_1.

The tree network design considered so far regarded the cost of each line as constant and equal for all links. As already explained, the cost of the link can affect the position of the Steiner point and must be taken into account when constructing the minimum network cost. The design stages, where the cost of the links is not equal, is the same as before, although the triangle which is circumscribed is no longer of equal sides. The relative cost of the lines between the Steiner point and the nodes is supposed to be known, as calculated in Figs. 7.18 and 7.19, and defined as $|AS| = P1$, $|BS| = P2$, and $|CS| = P3$. The triangle is then constructed with the base $|AB| = d$ and the sides $dP1/P3$ and $dP2/P3$, as shown by the example in Fig. 7.23.

Another two points to consider are the traffic volume flowing in the lines and the transmission lines available. The maximum transmission line capacity (i.e. speed) is of a fixed value. This value is not a function of the traffic volume that flows in the line, provided it is below the maximum possible capacity. The means whereby these points can assist in the design of the optimum network can be seen from the example shown in Fig. 7.24. In this example the network consists of three nodes ABC with peak traffic volume flowing between nodes $|AB|$ as 300 bits/s and between node $|AC|$ 950 bits/s, but with no traffic flowing between nodes $|BC|$. In this example only two types of transmission lines

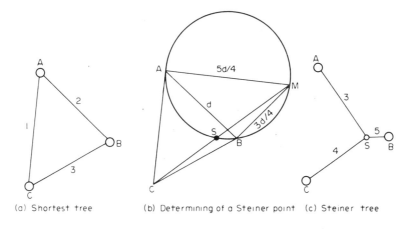

(a) Shortest tree   (b) Determining of a Steiner point   (c) Steiner tree

Fig. 7.23. Construction of a minimum Steiner tree with unequal branch costs.

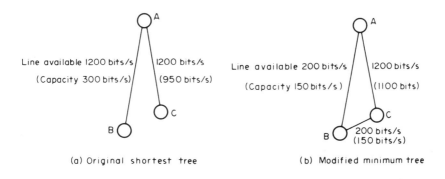

(a) Original shortest tree   (b) Modified minimum tree

Fig. 7.24. Split routing network structure.

are available, one capable of operating at 200 bits/s and the other at 1200 bits/s. Due to the traffic involved here, only the second type of line could be used, as shown in Fig. 7.24 (a). This network cannot be optimized by using the Steiner point. As the cost of both lines is equal, it will produce the relative cost $P1 = 2\,P2 = P3 = 1$, and this will not permit a triangle construction. Nevertheless, this network can be optimized by using other techniques, which are known as "split routing". If some of the traffic in the line connecting A and B nodes is diverted through node C, then link $|AB|$ could utilize the cheaper type line. That is to say, the original peak load of link $|AB|$ of 300 bits/s is divided into two streams, each of 150 bits/s peaks. Link $|AC|$ can still take the extra load, as it does not saturate it. The new split routine network shown in Fig. 7.24 (b) requires a new link to be installed between B and C nodes. This new network is more economical, as the links $|AB|$ and $|BC|$ use a much cheaper line than was originally used in link $|AB|$. The cost of line $|AC|$ remains as before, despite the increase in the load flowing in it.

Split routing is feasible provided the traffic flow in the lines is well below their saturation point. This technique is also based on the assumption that nodes B and C are intelligent enough so that they can relay the traffic through the new routings.

The cost of installing a number of channels along a single path depends also on the number of cables required. If the cable used can consist of N channels, then the cost of installing any number of channels where the total is less than N will be constant and independent of the number of channels required. This point could be used for optimizing the network by using split routing. An example where this minimization technique could be employed is shown in Fig. 7.25 (a). The example consists of three nodes ABC which are situated at the corners of an equilateral triangle. The traffic flowing in the network requires one channel in links $|AB|$ and $|AC|$ and thirteen channels in links $|BC|$. In this example the cables used may contain up to three channels in each. It thus demands one cable for each link $|AB|$ and $|AC|$ and five cables in link $|BC|$. The cost of the network may be reduced by diverting one of the channels from link $|BC|$ and routing it through node A, i.e. through links (AB) and (AC), as shown in Fig. 7.25 (b). The cost of links $|AB|$ and $|AC|$ will remain constant, though increasing the number of channels but not the number of cables, while the cost of the link between B and C nodes will be reduced by 20% by the removal of one of the original cables.

The optimization of the network shown in Fig. 7.25 can be continued by extending the algorithm to include a Steiner point. For this optimization, only the traffic on links $|AB|$ and $|AC|$ is considered, leaving out the traffic in link $|BC|$. This design aspect will produce a Steiner point for the traffic in the network, as shown in Fig. 7.25 (c). The traffic which was originally diverted from link BC to node A will be diverted only to the Steiner point. The new network, which uses both split routing and Steiner point, represents the most optimized network configuration possible for this example.

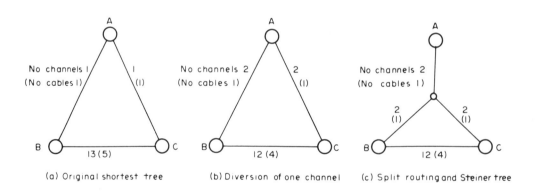

(a) Original shortest tree   (b) Diversion of one channel   (c) Split routing and Steiner tree

Fig. 7.25. Optimization of a tree network using both split routing and Steiner point techniques.

7.8 ALTERNATIVE ROUTINGS

In any distributed network with intelligent nodes, alternative routings are provided. Alternative routing is the ability to automatically select the best path to reach the destination and if this is not available to search for the second and third best routes. Alternative routings are required when the direct route is busy or faulty.

The design of any sophisticated network must consider the traffic volume which may be caused by alternative routing. If a particular link is designed with only the traffic volume flowing between its two nodes, the link will soon be overflowed and blocked. The design of any link must thus also consider the possibility that more traffic will be diverted to it from adjacent links when they become faulty. A contingency may arise where the traffic on each line is suddenly doubled or tripled. If this design constraint is inserted in each link in the network, it will cause the network cost to rise sharply, to a point where the whole network may not be practicable. The network will then be inefficient, consisting of many lines which are utilized well below their maximum capability. A compromise is therefore necessary so as to reduce this cautious requirement by allowing the messages to be detained longer in the system. The designer could accept overflow in the lines which may occur when certain links fail, with due consideration of the delays it creates. In other words, the designer can accept overflow in the lines when a failure occurs provided the diverted traffic will not cause the messages to remain in the system more than a specified time.

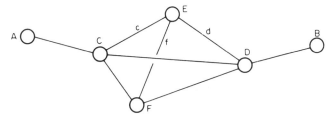

Fig. 7.26. Example network showing alternative routing between nodes A and B.

An example of a network with alternative routing is shown in Fig. 7.26. In this example there are five choices of routes for messages originating in a node A and destined to node B, but they can be divided into three groups according to the number of intermediate nodes on the way to their destination:

(a) Best direct route, two intermediate nodes, A-C-D-B,
(b) Second best route, three intermediate nodes, A-C-F-D-B,
$$A\text{-}C\text{-}E\text{-}D\text{-}B.$$
(c) Third best route, four intermediate nodes, A-C-F-E-D-B,
$$A\text{-}C\text{-}E\text{-}F\text{-}D\text{-}B.$$

The routing information is kept at the node in the computer store in the form of tables. Figure 7.27 shows a three-node network with the tables required at each node to provide direct and alternative routing. The number in the table states the number of intermediate links which the message must pass on the way to its destination. For example, a message in node B destined to node A and travelling on line b can reach its destination by passing two links.

Each table entry contains much more information than is shown in Fig. 7.27. In fact a separate line entry is required for each destination and for each node output link. The table should contain the following items:

(a) the destination address character of the remote node;
(b) line availability bit, 0 for line available and 1 for line unavailable;
(c) line output character number;
(d) number of links the message must pass on the way to its destination;
(e) final entry bit, 0, if another entry follows, 1, if this is the last line entry in the table.

TABLE 7.1

Dest.	Avail.	Line	No. link	Final
A	0	c	2	0
A	0	f	3	0
A	1	d	4	0
B	1	c	4	0
B	1	f	3	0
B	0	d	2	0
C	0	c	1	0
C	0	f	2	0
C	1	d	3	0
D	1	c	3	0
D	1	f	2	0
D	0	d	1	0
E	0	0	0	0
F	0	c	2	0
F	0	f	1	0
F	1	d	2	1

The reader must have realized that the address of the intermediate nodes is not given in this table, because this is of no interest at the originating node. Furthermore, the message, once it reaches an intermediate node, may be rediverted to another route than planned by the originating node. An example of such a routing table is given in Table 7.1 for node E in the network shown in Fig. 7.26.

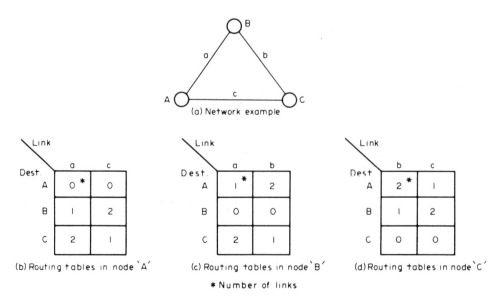

Fig. 7.27. Alternative routings-network and supporting tables.

A message arriving at the node will cause the processor to first examine the table locating the appropriate destination entries. Then the shortest route is located, i.e. the least number of nodes *en route* and not the shortest path length. Once a route is located it must be checked to ascertain whether the output links are available. If it is not available, another route must be located.

The whole operation is feasible provided the table is continually updated. This updating must take into account all the lines in the network and not only those originating from the particular node. In the table shown all the links were taken to be free except link |*DF*|, which was assumed to be faulty.

7.9 REFERENCES

1. Martin, J., *Teleprocessing Network Organisation*, Prentice-Hall, 1970, pp. 130-131, 153-169.
2. Martin, J., *System Analysis for Data Transmission*, Prentice-Hall, 1972.
3. Frank, H. and Chou, W., "Topological optimization of computer networks", *Proc. IEEE*, Vol. 60, No. 11 (Nov. 1972), pp. 1385-1397.
4. Gray, J.P., "Line control procedures", *Proc. IEEE*, Vol. 60, No. 11 (Nov. 1972), pp. 1301-1312.
5. Gilbert, E.N., "Minimum cost communication network", *Bell System Technical Journal*, Vol. 66, No. 9 (Nov. 1967), pp. 2209-2227.
6. Doll, D.R., "Topology and transmission rate consideration in the design of centralized computer communication networks", *IEEE Trans. Comm. Tech.*, Vol. COM-19, No. 3 (June 1971), pp. 339-344.
7. Karp, D. and Seroussi, S., "A communication interface for computer networks", *IEEE Trans. Comm. Tech.*, Vol. COM-20, No. 3 (June 1972), pp. 550-556.
8. Amstutz, S.R., "Distributed intelligence in data communications networks", *Computer*, Vol. 4, No. 6 (Nov./Dec. 1971), pp. 27-32.
9. Pyke, T.N. and Blanc, R.P., "Computer networking technology − a state of the art review", *Computer*, Vol. 6, No. 8 (Aug. 1973), pp. 13-19.
10. Ball, C.J., "Communication and the minicomputer", *Computer*, Vol. 4, No. 5 (Sept./Oct. 1971), pp. 13-21.
11. Tymann, B., "Computer to computer communications system", *Computer*, Vol. 6, No. 2 (Feb. 1973), pp. 27-31.
12. Roiz, E.F., "Considerations in the applications of modems to data processing systems", *Telecommunication* (Dec. 1968), pp. 19-23.
13. Hersch, P., "Data communications", *IEEE Spectrum* (Feb. 1971).
14. Doll, D.R., "Planning effective communication systems", *Data Processing Magazine* (Nov. 1970).
15. IBM: "General information − binary synchronous communications", Oct. 1970.

CHAPTER 8

Loop Transmission

8.1 INTRODUCTION

A completely new approach to the design of digital communication networks has been introduced in the last few years. This calls for the adoption of loop transmission to replace message switching for many applications. Both systems are intended as communication schemes for switching data between computers or between terminals and a computer. Where message switching serves large centralized centres with a need for active communication management, loop transmission serves random distributed network with little processing required.

This scheme is ideal for many Command and Control system applications where a number of computers operate independently in separate subsystems and are at the same time part of a comprehensive system, thus being able to communicate with each other to share the system's resources.

In the message-switching systems there is a highly sophisticated centralized control equipment which contains a computer. The initial cost is expensive, but not unduly so if distributed among all the potential customers of the system. Obviously, however, message-switching systems are not a practical solution for applications with a small number of customers, as the cost for each customer must be relatively high. Although message-switching systems are flexible, allowing many customers to be added after the system is in operation, it does have an upper limit at which the additional customers will reduce the serviceability of the system.

The loop-transmission communication system is superior in many respects to message-switching systems. In the first place the initial capital investment here is very low, as there is relatively small centralized equipment and the customers' terminals are simple. Then it is a most flexible system which can be constantly increased without affecting the general performance. New customers and new loops may be added without it changing any of the system's functions. Indeed the whole loop network may grow gradually, smoothly and economically.

In message switching, the initial investment is in the centralized equipment, while in loop transmission it is in the equipment for individual customers. Although the initial cost investment for each customer is relatively low, it may still be high with respect to each particular terminal. The cost effectiveness of the loop system is practicable only if the number of messages originating from each customer is relatively high, i.e. proportional to the line speed. Thus, for customers where the message output is small, such as sensors, keyboards and typewriters, the system may prove to be expensive if these terminals are directly connected to the loop. In these cases it pays to join a number of terminals together, using multiplexors or concentrators, and so increasing the accumulated number of messages that are fed into the loop from a single station. The multiplexors and concentrators could be inserted into the loop system as part of the area stations which feed the loop.

Another advantage which proves the superiority of the loop transmission is the relative volume of data that may be transferred in a single line using the same transmission speed. In multidrop lines, as used in message switching, each station is polled or selected by enquiring whether there is a message to transmit or whether the station is available to receive a message. Considerable time is thus required for setting up any message transfer. There is an average time loss of 50% for the dialogue between the message-switching centre and the area station when setting up the link. This time loss is estimated on the accumulation of the time it takes to transmit each polling enquiry, the station's answer and, above all, the turn-around time at the station and centre for changing the modem from the receive to the transmit mode. In the loop transmission system all these delays do not exist, and the proportion of real information messages is nearly double those transmitted in polled lines.

The speeds in the loops may vary from the standard data transmission rates up to extremely fast speeds. Bell Telephones have introduced special cables, termed T1, that may operate with speeds up to 1.5 meg bit/s, which have proved feasible for loop transmission. In the low-range transmission speeds, voice grade lines may be used adequately for the loop systems.

The loop transmission system is still in its introductory development stages. There are still a number of service problems which have not yet been solved, although this is presumably only a matter of time. An example of a service problem which has not yet fully been solved is the multiaddress message. If the first station in the address list deletes the message, how then will the others receive it? If it is not deleted, how will it be known that it was received, and if one station on the list is out of commission, how will the originator be informed about it? The only solution seen at this stage is to send separate messages to each destination, and have the sender's area station sophisticated enough to perform this automatically.

8.2 BASIC CONCEPTS OF THE LOOP TRANSMISSION

The loop network is constructed by means of a closed circuit which starts from a loop control unit, runs through a number of area stations and returns back to the loop control unit (as shown in Fig. 8.1). All the terminals are attached to the area station and from there transmitted into the loop. The data can flow in the loop only in a single direction. Any message originating at one point returns to the same point after a complete circulation of the loop.

The loop-system operation can be clearly understood if expounded with the aid of an example. A conveyor belt which has framed slots inserted in it is driven continuously in a circle, as shown in Fig. 8.2. At one point of the circle there is a control unit which pushes the conveyor belt along with a smooth continuous movement. The conveyor frame slots might contain packages for conveyance between stations. Whenever a frame passes a station the label on the package is checked to see if it is directed to that particular station. Any packages intended for that station are then taken off the belt and dispatched to their local destination. If a station has a package to send to another station, it just

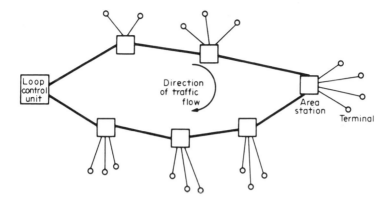

Fig. 8.1. Basic closed-loop transmission system.

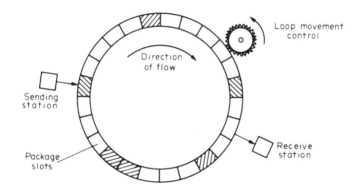

Fig. 8.2. Conveyor-belt delivery system.

waits for a vacant frame slot, dumps the package in it and the conveyor belt then transfers it to the other station.

The loop transmission operates in the same way as the conveyor belt. The operation is similar to a time division multiplex, with time slots made available for the transmission of messages. In the loop transmission system, message cell slots of fixed size are transmitted continuously around the loop. If a sending station has information to transmit, it looks for a free vacant cell, and when it finds one, puts the message in it. To ensure that the message that was placed in the loop will reach its destination, an address must be added to it, giving the name of the destination's area station and the name of the terminal in the area it is intended for. At the receiving station, every cell message address is inspected, to check whether it is intended for that particular station. If the address in the message tallies with that of the station, the message is lifted off the loop (freeing the cell) and then sent to the local terminal it is intended for.

The message cells are in a continuous movement in a single direction around the loop. The cell format is renewed each time it passes through the loop control unit. This ensures the cancellation of any delays or distortions created by the line characteristics. The size of the cell flowing in the loop can be different for each communication system; some systems may have cells with only a word length message, while others may contain cells with messages a block long. However, once a message size is selected for a particular communication system, it must remain fixed, with no possibility of changing it. In other words, the message size which consequently controls the cell size must remain fixed, because it influences all the hardware circuitry required for the area stations.

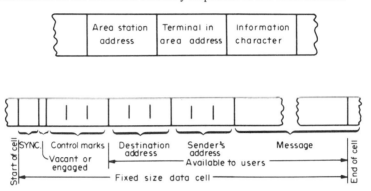

Fig. 8.3. Data cell fixed format.

The format of the fixed sized cells which consist of the message is shown in Fig. 8.3. Between the various cells a special synchronization code is transmitted to enable the stations to synchronize their clock and character start with that of the control unit. There is also a data bit, informing the stations whether the cell is vacant or engaged. Following that, there are a few control characters which are for the sole use of the control unit, to enable it to supervise the loop operation. The significance of these control characters will be explained later. They are for the loop operation only, and are deleted from the message when they are transferred from the area station to the terminal. Each cell must contain both the destination address, the message it is intended for and the sender's address. The address format is divided into three parts: the area station address, the terminal address in the area, and a message character containing information identifying the message. Only after all the peripheral data is transmitted can the message itself be sent on.

All the stations have equal importance, with no special duties attached to any particular cell in the loop. No priority scale, as known in message switching systems, is used in the loop transmission system. Thus all the area stations have an equal chance of reaching a vacant cell to transmit a message. The loop system operation may be regarded as the random seizure of a cell. This improves the efficiency of the system compared with the use of time or frequency division multiplexing, where there are time or frequency slots allocated to each terminal which are not fully used. The same applies when comparing the loop system with the polling systems (as already explained).

Theoretically, each time a cell passes an area station it must wait for the station to check the address and to lift the message from the loop. This may cause delays, which will increase as more area stations are added to the loop. This problem is easily overcome by arranging the logical speed of operation of the circuits in the area station to be much faster than the flow speed of information transmission in the loop. In this way the whole operation may be concluded before the cell leaves the station.

In a system which operates in a random manner, there is a danger that a single area station will completely dominate the whole loop by inserting messages into it continuously without leaving any vacant cell for other area stations. This random operation will soon reach a saturation point, thereby blocking the whole loop. This problem is the more critical because of the fact that the area station operates faster than the loop transmission flow. To avoid this problem the area stations must be designed so that they are limited in the number of messages they can transmit in a given period of time. This is achieved by holding up the area station and preventing it from transmitting a second message immediately after a first message has been transmitted. The delay factor is calculated by a statistical analysis of the messages in the loop, which must be considerably smaller than the loop transmission speed. Although the area station must wait between the seizure of cells, it must check each cell that passes to see if there are no messages intended for it.

In any communication system there is a danger that messages will not reach their destination, or that they are lost in the system. This could come about when the area station, sending or receiving, is busy, is faulty, or out of commission. In the design of the loop transmission system, some means must be inserted to ensure the safe transmission of the messages, as will be explained in the following section.

8.3 LOOP SYSTEM OPERATION

As previously explained, every transmission loop consists of a control unit which circulates the information continuously round the loop. The number of area stations inserted in the loop may be as many as required; there are systems with as many as 10, 50 or even 100 area stations. The number of stations is, however, limited by the loop speed and the message load in the loop. If the traffic is increased beyond the capacity of the loop, then another loop may be constructed, as will be shown later.

The control unit of the loop has been labelled 'A' station, and the area stations as 'B' stations (as shown in Fig. 8.4). The characteristics of both stations are discussed below:

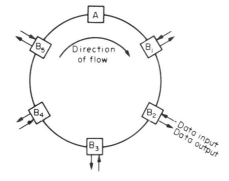

Fig. 8.4. Basic closed-loop transmission network.

'A' station

The control loop station, shown in Fig. 8.5, has the task of supervising all the operations of the loop and ensuring that no operation errors occur. It has two main

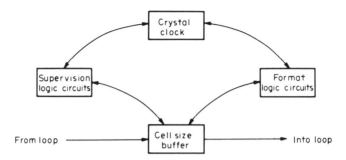

Fig. 8.5. "A" station logical diagram.

functions: renewing the message format by ensuring that synchronization is maintained, and supervising the loop performance by preventing the build up of traffic in the loop.

Station A renews the cell format each time it passes through it, creating a full message delay. For this operation a full cell size buffer is provided. Each message is lifted off the line and then returned with the right timing. In this way the format is reshaped and reproduced, disregarding whether there is information in the cell. The control station consists of a crystal clock which enables the cell to be reproduced accurately and to synchronize the operation of the whole loop.

The second function of station A is to be on guard so that the loop is not saturated with messages that have not reached their destination; in other words, to ensure that the messages do not circulate in the loop indefinitely. The first time a message passes through the A station, a marker in the form of bit '1' is placed in the control character to indicate the fact. The second time the message passes through station A, it can then be assumed that the message did not reach its destination and in fact has been lost in the loop. It must be realized that in this system the B station can never be busy. If a B station does not lift off a message intended for it, it can be assumed that the particular B station is out of commission or that there is an error in the address part of the message. Thus, if B station does not lift the message the first time it circulates the loop, it will not do so on a second turn.

Station A, on finding that the circulating message has not reached its destination, may destroy it by erasing the information, thereby liberating the cell. A better solution is for station A to return the message to the sender, accordingly informing him that the message has not been delivered or accepted by the destination station. Station A performs this operation by simply interchanging the position of the sender's address with that of the destination. In this way the message is returned to the originator, who can then decide what other means to adopt for the dispatch of the message.

The interchange of addresses when the message is not delivered solves the problem of whether the destination is out of commission. If the fault lies in the sender's equipment, there is a danger that the message will not be accepted back, which creates the danger that the message will again circulate indefinitely in the loop. To meet this contingency, the cell must be marked by station A in the control character to indicate that the addresses have been interchanged, and the second time the marked cell passes A station, the information must be destroyed.

The A station must keep complete track of all the messages that do not reach their destination, and must inform the supervisor of any transmission faults so that he can act accordingly. It also keeps a record of all the B stations where the faulty message originated from or to where the message is destined. A warning will be given if the number of messages to or from a B station have not been handled properly, this being a good indication that there is an error or fault in the B station itself. If this is the case, all the other stations in the system must be warned not to send any messages to that particular station.

Most of the faults in the B station operations could be detected by the A station by checking the cells that pass through it. However, there are some faults which cannot be realized by the methods described. The main problem arises when B station writes a message in the loop but is not capable of marking the cell as being engaged. This will cause the message to be left undetected in the system. This situation will not saturate the loop, as the cell remains free and new messages can be inserted in it by other stations. The information, however, is lost in the system, with no means of it being traced. To avoid this problem, A station must send test messages to each B station in turn every given period, and the B station must respond so that all the control and message bits accessible to the area station are checked for a fault. A special cell engaging code must be used by the check format to distinguish it from the busy/free marker which is operated by the B station. The check format consists of the B station address and 0's in all the rest of the cell. The B station then returns 1's, which indicate the proper operation of each bit.

'B' station

The B stations are the interface between the terminals and the loop. Their object is to write messages into a cell or to lift off the messages already in the loop, addressed to the station. As already stated, B stations must always be accessible for messages addressed to it.

The B station must synchronize its operation with that of the A station. This is performed with the aid of synchronization signals transmitted in the front of each cell. These coded pulses enable the local pulse generator to synchronize its operations with that of the loop not only with regard to the bit synchronization but also with the character and the start of the cell format.

Although the B station must never be 'busy' but always available to receive messages from the loop, the loop itself may be busy and the messages arriving from the terminals may then have to wait in a queue in front of the B station till a free cell is found to transmit them. A buffer should be constructed into the B stations to store the messages till they can be transmitted (as shown in Fig. 8.6). The size of the buffer must be

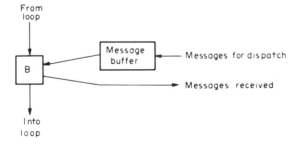

Fig. 8.6. "B" station general configuration.

calculated by a statistical analysis of the queue of messages that is created by the random number of messages to be transmitted and the random chance of finding a free vacant cell. In the logic of the buffer, a priority scale may be added to cope with the local high priority messages.

The operation of the B station is as follows: each and every cell that passes the B station is checked to ascertain whether it is vacant or busy. Two marker bits need to be referred to, because the first could have been inserted by another B station showing that a message has been placed, and the second could have been inserted by the A station to check the station operation. If the cell shows a busy sign, it checks the address of the

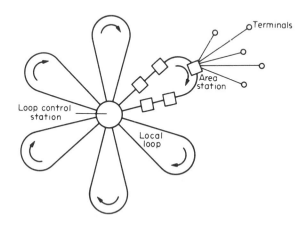

Fig. 8.7. Centralized loop network.

message to see whether it is intended for that particular station. If the cell shows a vacant sign, it can then insert a new message to be transmitted.

When B station sends a message in the vacant cell, it must also perform the following parallel operations:

(a) Raise the 'engage' flag in the cell by interchanging the marker from 'vacant' to 'busy'.

(b) Add the required address consisting of particulars of both the originator and destination.

(c) Insert the complete message.

(d) Inform the local terminal, or the host computer from where the message originated, that the message has been dispatched. This is arranged so that more messages can be fed into the B station from its associated terminals.

When B station receives a message which is addressed to it, it must perform the following parallel operations:

(a) Inform the computer, with an interrupt signal, to accept the message.

(b) Transfer the message from the loop to the computer or terminal, directly or via a buffer (only the message itself is transferred without any of the marker and control signal codes).

(c) Cancel all the information in the cell which is accessible to the B station, that is, both the information and control characters.

(d) Change the marker in the cell from busy to vacant.

(e) Erase all markings inserted by the A station.

When A station checks the operation of the B stations it sends a special coded marker in the control character which is accessible only to the A station. It also adds the address of the B station which is to be checked. The special code informs all the other B stations in the loop that the cell is engaged. The B station which is checked changes the polarity of all the bits in the cell which it may have access to, both in the message and the control part of the format. In this way the A station may check the validity of all the B station functions. The engage marker code inserted by the A station must be distinguished from that which could be inserted by the B station. This permits checking the full operation of the B station, and in particular the function of inserting a busy marker when the station seizes a cell.

All the B stations in the loop are identical and have equal importance in the operation of the loop. They may, however, have extra duties in respect to the terminals attached to them, serving possibly as concentrators or communication controllers buffering the messages which arrive randomly from the various terminals.

8.4 LOOP NETWORKS

The basic loop configuration discussed in the previous section can be expanded into a large sophisticated network. The efficiency of the whole system depends on the potentiality of its being expanded to include a number of loops which may be interconnected. The first suggestion (as shown in Fig. 8.7) interconnects all the loops at one point. All the messages go to or from a central processor, similarly to the message-switching system of a star network. This system does not seem practical for a

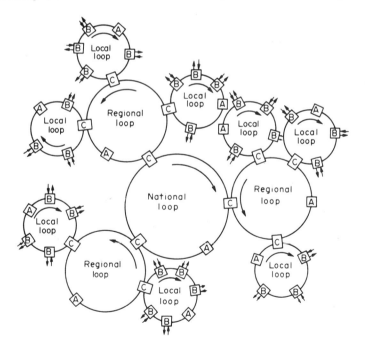

Fig. 8.8. Hierarchy of loop network.

multi-user system spread over a large geographical area. Efficient operation can be achieved by building trunk loops which are identical to the basic loop and are used to interconnect the various loops (as shown in Fig. 8.8).

Each loop in the proposed network may be operated independently and at its own transmission speeds. Accordingly, an A station must be inserted in each loop, to control the operation speed of that particular loop and supervise the functional performance of the loop. Messages may pass from loop to loop through special interconnecting stations, termed C stations.

All the loops could function equally with the B stations inserted in each of them, and then the messages could flow through a number of equal loops till they reach their destination. Alternatively, two types of loops might be used, local loops with their B stations and trunk loops. This would simplify the address format. In large networks where the stations are scattered over large areas, a hierarchy of loops may be used. In Fig. 8.8, a three-level hierarchy system was shown, having two of the levels as trunk loops. Nevertheless, a hierarchy of a more complex network is possible. The top level is intended for the national loop, with speeds up to 1.5 meg bit/s. There are also the regional loops with speeds up to 50 kbits/s and the lowest level with loops for local operation which may use speeds as low as 2.4-4.8 kbits/s.

The C station is planned to function as an intermediate station between two loops to enable the transfer of messages from one loop to another. When considering a single loop in the network, the C station behaves in an identical manner to that of the B station. That is, it lifts messages from the loop addressed to it and inserts messages into the loop originating from other loops. For this purpose the C station is constructed of two B stations arranged back to back, as shown in Fig. 8.9. Buffers must be inserted between the two B stations, since messages must be queued till a vacant cell is available. The size of the buffer here may have to be larger than that in B stations, as the C stations may also have to convert the speed from one loop to another.

It is technically feasible to add terminals and computers to the C station, and then it could be operated not only as a loop interchange station but also as an area station. Where this applies, special extra logic must be added to the C stations so that the messages addressed to this particular C station are not transferred to other loops.

The large loop network requires a new type of message format, because extra address information must be added to the cell to include the trunk loop address. The C station checks every cell that passes the station, without delaying the message flow. When

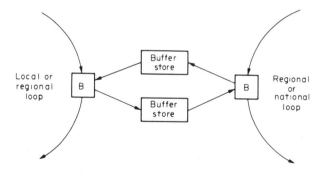

Fig. 8.9. General configuration of "C" station.

transferring messages from the local loop to the regional, or from the regional to the national, the C station only checks the regional address of the cell format. If the address is different from that of the original loop, the station lifts it off one loop, and thereby transfers it to the other. On the other hand, when transferring from the national loop to the regional loop, the regional address is considered, while when transferring from the regional loop to the local loop, it is the local address that is considered. In both these cases, the address must be identical with that of the destination loop.

When the C station transfers a message from loop to loop, it transfers only the message and the address information but it does not transfer any of the control characters, since this data is specific to a particular loop and its associated A station. Transferring the control character would create disorder in the operation, that is, having messages erased by the new A station after they have been marked by the original A station of the first loop.

When the buffer storage in the C station is full, due to an unexpected peak volume, all the B stations in the loop must be informed. The area B stations may then wait until the passage is confirmed, or it may look for alternative routes or alternative destinations. The system should be designed so that this contingency is obviated. This could be achieved by increasing the buffer storage in the C station and by inserting alternative routes.

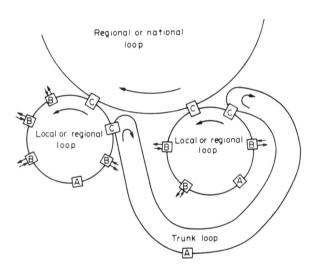

Fig. 8.10. Special trunk loop for diverting traffic of a local nature between adjacent areas (loops).

8.5 ALTERNATIVE ROUTES

Alternative routes are required both when the selected route is overloaded or when it is faulty. This calls for broadly two types of alternative routes, a trunking loop to divert the traffic from an overloaded loop and alternative loops which operate when the normal loop fails.

A large percentage of the traffic in the regional (or the national) loops may be accounted for by messages communicating between adjacent local (or regional) loops. This traffic is mostly of a local nature, but by loading the regional loop, it can prevent

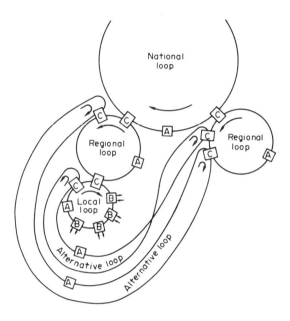

Fig. 8.11. Alternative routes.

long-distant messages from entering the loop. To prevent the heavy local load from monopolizing the regional loop (and its associated C station) a special trunk loop could be constructed between two adjacent local loops, as shown in Fig. 8.10. The special trunk loop acts in the same manner as the standard regional loop except that it has only local loops attached to it, with no access to the national loop. The trunk loop must have an A station, to ensure proper loop operation with it as with an ordinary loop. In this scheme, each local loop has two C stations, one connecting to the regional loop and the other to the trunk loop. The two C stations of the local loop operate in harmony; both may accept messages to be transferred between the two adjacent loops, although one of the C stations must also accept messages for other destinations. In order that the new C station will take the full bulk of the traffic load between the two adjacent local loops, this station must be positioned in the loop before the C station which connects the local loop with the regional loop. Any messages to other local loops will be transferred to the regional loop, while all the messages between the two adjacent loops will be lifted before they reach the regional loop and will be transferred to the trunk loop. The two C stations in the local loop must be situated relatively close, with no B stations between them, so that all the messages may be first intercepted to see if they are eligible for the trunk loop.

The same trunk-loop system connecting two local loop pairs could adequately be applied between two regional loop pairs.

The trunk loop is intended to divert the traffic load from a saturated loop to a substitute route. There is also a need to create alternative routes which come into operation if the normal path is faulty. Figure 8.11 shows two types of alternative routes. The first is the interconnection of two regional loops, which comes into operation when the national loop is faulty. The second connects a local loop with another regional loop which comes into operation when the first regional loop is faulty. These alternative loops must also consist of an A station in each, and are connected to the other loops by means

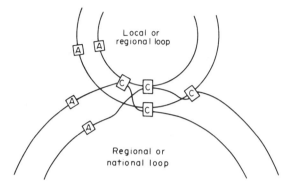

Fig. 8.12. Double-loop operation.

of C stations. Here, however, the new C station is placed after the first C station. In this case the traffic is through the standard loops, whereas the alternate loop transfers messages only when the normal routes are defective.

Reliability may be increased by duplicating all the loops and the interconnecting C stations, as shown in Fig. 8.12. This system should be operated on a shared-load basis, where one loop takes the full load when there is a fault in the other loop. Every message may be placed in either of the two loops, but not in both of them. This is because the A station operation is separate for the two parallel loops. Thus, there could be confusion in the system operation if in one of the parallel loops a message is dispatched back to the sender while in the other parallel loop the message has already been delivered. The actual operation must transfer each message only once and in only one of the parallel loops. As shown in Fig. 8.12, it transfers the messages from loop to loop via the first C station it comes across. If that particular C station or its associated loop is out of commission, it continues in the original loop till it reaches the second C station.

The discussion has concentrated on the alternative routes when there is a fault in a loop or in an interconnecting station. This is not the only contingency to be considered, for a whole system may be put out of commission if the B or A stations are faulty. In such an event the B station must be switched out of the loop the moment it does not function properly, as shown in Fig. 8.13. This will allow the loop to continue its operation; nevertheless, all the messages addressed to the deleted stations are returned to

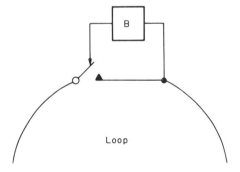

Fig. 8.13. Switching in/out "B" stations.

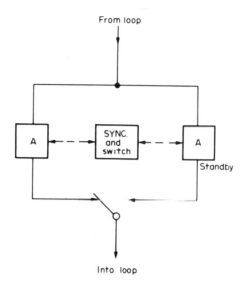

Fig. 8.14. Switching standby "A" station.

the sender. This solution cannot be effective in the case of a fault in the A station, as that is the heart of the loop operation. Figure 8.12 showed the double-loop solution which is also effective when one of the A stations in the parallel loops is faulty. However, the importance of the A station in the loop calls for higher reliability, since a faulty A station takes the whole loop out of commission, including all its associated B stations. This being the case, a redundancy system is called for. Two A stations may be operated in parallel, one of them as a stand-by station which is switched in when the first station is faulty (as shown in Fig. 8.14). Both these parallel A stations must be continuously synchronized. Also, the switch-over between the two stations must require negligible time, so that no messages that are already in the loop are lost. The standby A station must be fully functional, performing all the operations parallel with the first station, although only one of the two A stations can supervise the loop operation at any given time. These requirements are made feasible by connecting the inputs of both A stations in parallel to the loop, with the output of only one of the two A stations connected to the loop.

8.6 REFERENCES

1. Martin, James, *Teleprocessing Network Organisation*, Englewood Cliffs, N.J., Prentice Hall Inc., 1970, pp. 122-125.
2. Pierce, J.R., Coker, C.H. and Kropfl, W.J., "An experiment in addressed block data transmission around a loop", *IEEE International Convention Record* (Mar. 1971), pp. 222-223.
3. Hays, J.F. and Sherman, D.N., "Traffic analysis of a ring switched data transmission system", *Bell System Technical Journal*, Vol. 50, No. 9 (Nov. 1971), pp. 2947-2978.
4. Avi-Itzhak, B., "Heavy traffic characteristics of a circular data network", *Bell Syst. Tech. J.* Vol. 50, No. 8 (Oct. 1971), pp. 2521-2549.
5. Pierce, J.R., "How far can data loops go", *IEEE Trans. Comm.*, Vol. COM-20, No. 3 (June 1972), pp. 527-536.
6. West, L.P., "Loop-transmission control structures", *IEEE Trans. Comm.*, Vol. COM-20, No. 3 (June 1972), pp. 531-539.
7. Anderson, R.R., Hayes, J.F. and Sherman, D.N., "Simulated performance of a ring-switched data network", *IEEE Trans. Comm.*, Vol. COM-20, No. 3, pp. 576-591.

8. Spragins, J.D., "Loop transmission systems – mean value analysis", *IEEE Trans. Comm.*, Vol. COM-20, No. 3, pp. 576-591.
9. Pierce, J.R., "Network for block switching of data", *Bell Syst. Tech. J.*, Vol. 51, No. 6 (July-Aug. 1972), pp. 1133-1145.
10. Kropfl, W.J., "An experimental data block switching system", *Bell Syst. Tech. J.*, No. 6 (July-Aug. 1972), pp. 1147-1165.

CHAPTER 9

Computers in Command and Control Systems

9.1 INTRODUCTION

As already defined the Command and Control system is organized in a closed loop where data continuously flows up and down the line. Computers are employed both in the command and in the data collection centres. Decisions for execution are sent forward and the results are continuously fed back, all processes operating in real-time. The best control of the organization resources can be effective only if the transmission of the information in both directions is fully exploited.

It is a basic requirement of a Command and Control system that the computers involved are communication orientated and can be operated on-line in a real-time environment. The adopted computers must possess the following qualifications: (a) They should be capable of handling masses of data swiftly and efficiently and store large quantities of information. (b) They must also be able to operate a large number of input enquiries simultaneously and respond to them expeditiously. This chapter outlines and discusses the computer configurations which may be obtained in Command and Control systems.

The computer configurations available are part of a fast changing art. There are many computer combinations which are effective but fail to present a satisfactory cost performance. As the system grows in scale and complexity, the problem of understanding and predicting system behaviour becomes increasingly important. It is essential ever to bear in mind that the computers are not a main objective of the system design but only a means of solving particular processing problems. The key problems to be studied are the interfacing of the computer with the large number of data inputs and the sharing of the computer resources by all the inputs.

Command and Control system designers are expected to develop a complete system configuration plan to include provision both for the collection of distributed data and for the central processing of the information. Ideally the communication system designer should be also an expert on computers, but, generally, communication engineers are not proficient in computer technology, and conversely the computer engineer has inadequate mastery of the subject of communication. It is customary to resolve this difficulty by organizing a separate associated team to deal with all the software problems and to have a constant exchange of design information between the computer designers and the communication system experts. However, the system expert must be involved in the design of the computer configuration, since he must still be responsible for developing, recommending and executing the system design operational procedures. A system designer cannot demarcate the point where his responsibility ends and that of the computer designer begins. While the communication expert is not expected to design the computer software, he must be capable of defining the computer configuration needed for the system and later be able to evaluate the proposals presented to him. In other words, the system designer is solely responsible for the specification, which is to be based on two design points:

(a) Availability, that is, specifying the response time required to execute each enquiry.

(b) Reliability, that is, specifying what and where redundancy can be introduced into the system.

An example of these design requirements was shown in Chapter 6, where duplex message switching was introduced. The configuration there was based on an active standby computer used to ensure continuity of service in case of system component failure. The system designer does not complete his activity by defining the computer configuration. He must, for instance, have to decide whether a duplex system is to be used, whether each half system operates on an equally shared load basis or on a standby basis. It is the system designer and not the computer designer who must elaborate the configuration, to show how the system will be split, what activities each half will perform, whether the common data base will be shared, how much of the data base should be common and how it will be accessible. The computer engineer in turn is the one who will define the data base operation, but this of course only after the system designer has specified the configuration and operation. The system designer and the computer engineer may have conflicting views, since they regard the system from different approaches. However, it is crucial that the design approaches be made conjointly. Only after the configuration of the complete system has been specified, including the remote terminals and the intermediate centres, can the design of the computer software be carried out. The performance evaluation of the computer workload will then be based on analytical moduling, on simulation and on the measurements of the problems in question.

The computers for the ultimate adoption can be finally defined only after the following aspects have been considered:

(a) How completely are the communication lines taken into account?

(b) What work load is the system to perform?

(c) What part of the workload must be performed in real-time?

(d) What processing is required in real-time and what can be delayed?

(e) How is utilization of the processor power distributed among the transaction modules?

(f) How much processor time is required for specific requirements?

(g) What is the data base to be constructed?

(h) What type of enquiries are allowed in the system and the maximum time for an answer to be given?

(i) How do the answers to specific enquiries vary from customer to customer?

(j) What data should be updated, at what frequency, and by whom?

(k) What priorities are to be used in the system?

(l) What privacy precautions are to be introduced into the system to safeguard against unauthorized people reading or changing data?

(m) What computer language is to be used in the system?

In this chapter the software aspects of the computer will not be discussed. What will be given is a general outline of various computer configurations that may be used in

Command and Control. In particular, the subject of time-sharing computer organizations and the various aspects of computer hierarchies will be examined.

9.2 TIME SHARING

The most important feature of Command and Control systems is that it enables many users to communicate with a central processor where the connection is on-line and the operations are performed in real-time. Although many users are connected to the system and operate simultaneously, each user will appear to have the computer exclusively to himself. This is effected by sharing the computer resources in time division and providing direct access to the user so that he may be able to carry on a dialogue with the computer. Decision making, which is the main task of Command and Control systems, can be performed only if the information is collected from many sources simultaneously and then displayed in real-time.

Time-sharing computers permit a large number of users to operate quickly with direct access to the computer at the same time, as shown in Fig. 9.1. Here each user is able to enter a programme at any time and from any terminal, and can expect the computer to react and reply instantaneously. This is possible because the terminal is provided with a direct communication link with the central processor.

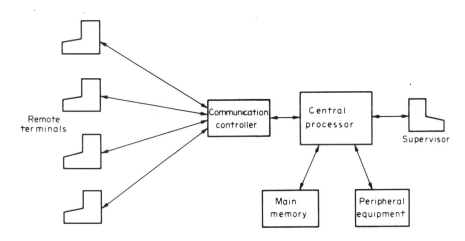

Fig. 9.1. Basic configuration of a time-sharing system.

In batch processing, all the programmes are fed separately and are operated in sequence. In consequence, the user may have to wait hours or even days for the computer reaction. Furthermore, the computer has to run a complete programme without stopping and this usually causes the other programmes to wait long periods before gaining access to the computer. If there is a bug in the programme, the computer will automatically throw out the programme until the mistake is corrected, and the programme is returned to the end of the queue. These problems do not arise in time sharing systems. Here the user views the computer as exclusively his, and is not bothered by other users or by queues or delays. In time-sharing systems there could be a continuous dialogue between the computer and the terminals even though they might be miles apart. The user must get an immediate reply to all his requests, and if there is a fault (bug) in the programme, the user will not be excluded from the system but will get an immediate notification. Furthermore, the user is able to correct the fault without having to feed the whole programme again and can continue with only negligible delay. Above all, the user can insert changes during the dialogue, or modify the programme as he wishes.

The above comparison should not imply that time sharing is intended to replace batch processing or that it is invariably more efficient and more economical. In fact, time-sharing systems are more expensive (although they present better efficiency to the individual user) than batch processing. Each of these processing techniques is intended for specific applications. In time sharing, much of the processor time is devoted to relative non-productive functions (such as programme swapping or turn-around time) and thus it is less cost-effective. It is intended essentially for short programmes which require real-time problem solving to save programmers' man-hours. Large programmes which require long processing times are generally performed by batch processing. Thus, for most scientific applications, where many calculations are required, batch processing is the more economical; however, for Command and Control systems, time-sharing techniques are essential, as without them the whole system may not be operative. It can be said that the development of time sharing is the main factor that has determined the advancement of Command and Control systems. Despite the high cost and the requirement for a larger computer configuration, Command and Control cannot be feasible without the sharing of its resources.

Time sharing makes demands on both software and hardware. It requires a large computer configuration to enable a number of consecutive programmes to operate simultaneously at a high speed. Basically it must be fast enough to respond to any task within a required time interval. This time interval must make allowance for the full user programme to be processed even if it is segmented, without the user realizing that a delay has elapsed between his request and the computer response.

Although time-sharing computers are said to perform several different tasks simultaneously, this is not strictly true. It is obvious that a single processor can perform only one task at a time. What actually happens is that the programmes are run not simultaneously but in sequence. The intervals between the consecutive programmes will, however, not be felt by the user-operator, because the slow reaction of the human operator in conjunction with the high speed of the computer will give the operator the feeling that the whole system is completely devoted to solving his tasks. Each programme segment will run for a short period of time, after which the next consecutive programme has its run. Thus all the programme segments will run in sequence for a fixed time duration, as in time division multiplexors. By using a multiprocessing configuration (as will be explained later) a number of simultaneous programme operations may be run in parallel, each programme executed by an independent processor.

In principle any large computer may operate in a time-sharing mode, although this is not always practical, as some changes are required in the computer. The basic requirements of a time-sharing computer are that it must always have a large main and backing memory capacity and that the cycle time will be as short as possible. In order that the computer may act efficiently in a time-sharing mode, two operational points are required:

(a) Re-location of the programmes, which permits the programme segments being processed to be located at different places in the memory each time after the computer completes a process period. The handling of the programme is complicated enough to justify a hardware rather than a software solution. Special address tables are used which may be dynamically changed each time the programme is returned to the backing memory.

(b) Zoning memory locations, which are essential when there is more than one programme in the memory at any given time. The zoning controls the circumference of area given to each programme instructions and its relevant data.

The difficulty in designing a time-sharing system lies in the demands on the supervisory programmes which are to control the whole system operation. Many programmes may be kept simultaneously in the main memory, and where multiprocessing is used, they may be processed simultaneously. The supervisory programme must provide processing time and memory locations which may be dynamically allocated and continuously changed. The users' programmes must be segmented into separate pages, which could be processed page by page. Thus the functions which the supervisory programme must be able to perform are as follows:

(a) Dynamic memory allocations.
(b) Programme segmentation into pages.
(c) The re-location of the programme pages in the memory.
(d) Job scheduling.
(e) Sharing of the computer resources.

As already explained, time sharing suffers from the fact that much of the processing time is devoted to relatively non-productive functions. For this reason the most important aspect of time-sharing system design is the efficient manipulation of the processing time. Hence, time sharing tends to be classified according to processor utilization. The utilization of the processors is a function of the job scheduling in the system and is based on the following features:

(a) Queueing. The processor deals with the programmes according to the programme priority. Once the processor handles a specific programme, it will exploit all the system's resources.

(b) Priority scale unit, which is assigned to each programme as it enters the system. The processor will always handle the highest priorities before it will deal with lower programme priority.

(c) Quantum time, which is the time allocated to each page process, i.e. this is the time each programme is allowed to remain in the processing facility. The duration may vary in size for each priority.

(d) Time accumulation, which is an algorithm recognizing that the complete programme is allocated a specific reasonable time for processing. This is an accumulation of all the times the programme is run on the processor. With short and medium programmes this maximum time is not reached, but for long programmes the time required may expire. Then the programme in execution is forced to terminate its operation so as to allow other programmes to be processed. The programme is then sent to the end of the queue and will be continued only when its turn comes again. The system must keep a record of all the programmes, indicating the various stages of processing, and recording the time and resources allocated to it so far.

(e) Natural wait, which is the particular time a job must remain without provision for entering the processor. This waiting time could be caused by an input–output operation. In most programmes this function is already introduced as part of its sequence operations. When the programme reaches a point where a natural wait is required, the processing automatically stops and the next consecutive programme segments are processed.

(f) Swap time, which is the time required to set up the necessary files of the departing programme and the time required to bring the new programme into the main memory.

(g) Demand, which is a request for a task to be performed by the processor.

These aspects will be expanded in the following sections dealing with memory hierarchy, queueing, priority and scheduling.

9.3 JOB SCHEDULING

The time-sharing process requires a supervisory programme which will act like an internal traffic cop, controlling signals to the various programme segments entering the main memory from the backing memory. This allocation of the resources is performed by a job scheduling, which first identifies, then handles each programme in turn and prevents the interference of conflicting demands.

The use of many terminals which operate in parallel and require simultaneous handling by the centre processor will create a queue of programmes waiting to be processed. Scheduling is intended to handle the queues and thus increase system efficiency. Each programme is divided into pages of average working size. To process each page, the page is given a quantum time slice which is allotted so that the page process becomes inactive before the running out of the time slot period. The allocation of the time slices and the selection of the pages is performed by the scheduler.

The efficiency of the system can be increased if the response time of short jobs is reduced. The real processing time of a page process is the sum of the actual process time, the swap time, the natural wait time and the time wasted on programme faults. The response time of each process should not be too short, as it may result in the short programmes having too many runs; on the other hand, they must not be too long, so as to give all the programmes an even chance to run. The time selected should be greater than the average actual process time of short jobs, to allow for a possible small number of programme faults.

Another means of increasing the system efficiency is the introduction of a priority scale for the programmes. This priority is internal and could be defined by the programme length or other processing functions. The priority scale should also be used to handle any conflicting requirements of the users by placing each programme in the queue according to its relative importance. Although a priority system is used, the scheduler must ensure that equal rights are presented to each programme. There is always a danger that the processor will only handle high-priority programmes and neglect the lower ones. One method to avoid this is by means of the accumulation time and the quantum time.

Fig. 9.2. Basic queue model which deals with the programmes according to their arrivals.

In batch processing the programmes are handled consecutively, commencing with the first programme to arrive. The programmes remain in the system for the full period required to process the programme irrespective of its length. Here a simple queue of programmes is created in the system (as shown in Fig. 9.2). In time-sharing systems, however, the simple queueing operation is not practical, as this would cause long delays, with one terminal (programme) possibly dominating and preventing the programmes of other terminals entering the system. In these circumstances each programme has to be interrupted to allow other programmes to enter the system, despite the fact that the first programme has not completed its programme tasks. Here the central processor is presented with a queue which is attended to according to various scheduling algorithms. The scheduling allocates a finite service to each successive programme and then returns the uncompleted programme back to the end of the queue. A programme may be interrupted when the allocated quantum time has ceased or a higher priority programme is waiting. The uncompleted processed programme will then return to the queue and will be handled when its turn comes up again. In this way each programme may be interrupted a number of times and have a number of processing runs. Even high-priority programmes may be interrupted if they overrun the quantum time allotted to them. However, they will always have a priority to advance them in the queue. When the programme processing is completed, the programme will be ejected from the system. This may occur once or any number of times after the programme has entered the queue.

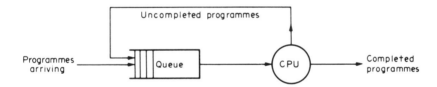

Fig. 9.3. Basic queue model for a time-sharing system with programme slicing and feed-back of uncompleted programmes.

In short, the whole efficiency of the system is based on the scheduling algorithm. Its functions are according to the priority allocated to each programme and to the ability of the scheduler to assign both the quantum and accumulative times. One of the features of the scheduler is that the processor is never unemployed when programmes are waiting in the queue.

The queueing model used for time sharing is referred to as feedback queueing, since each programme, when interrupted, is fed back to the end of the queue (as shown in Fig. 9.3). The queue itself handles the programme slices one after the other on a first-come-first-served principle. (With a priority scheme the programme slices are allowed to jump the queue, i.e. advancing their position in the queue, if they are at a high priority.) This is a well-known scheduling procedure, known as the Round Robin discipline, where each programme is assigned a fixed quantum and each job thus receives the maximum time and then returns to the end of queue to await another quantum of service. The priority in this scheme is based on the time of arrival in the queue and the processing time used within the quantum time. This, however, reveals the main disadvantage of the scheme, as the priority can be defined only after the programme has entered the system.

As already stated, the aim of time-sharing scheduling is to provide each user with the feeling that the computer is completely reserved for himself. Thus when he has a short enquiry he will expect a fast reply but when he has a long processing problem he will understand that his demands will take time. This presents one of the basic scheduling algorithms, where short programme requests will get the highest priority and the longest programme will be given the lowest priority. This will allow the processor to deal quickly with short programmes and fit the long programme into the spare time left.

When a programme has to be halted for any reason, such as when it must be given a natural waiting time till further data is available, the programme slice with a high priority must not enter the queue, as it will interfere with the routine operation. The priority can be defined, if possible, by the time the further data is available for continuation of processing.

Another aspect that could be used for defining the priority is the programme requirement for special system resources, such as printing. The slower the operation of the required resources, the lower the priority given.

An important scheduling algorithm is to make sure that short programmes or the high priority programmes will not dominate the whole processing time, for this would create a long queue of low-priority programme slices with no chance of entering the processor. The system must be able to measure the waiting time of each programme in the queue and the accumulated waiting times between runs. If this time period is beyond the fixed time allocated to the specific priority programme, then the priority may be changed or a demand may be made to advance the programme.

The scheduling algorithms discussed so far may not always satisfy all the system time-sharing customers when their long programmes are only slowly processed. In Command and Control systems long complicated calculations may require faster response than shorter programmes, for example, in air traffic control, where the calculation of the plane location is of a higher priority than answering routine enquiries. In consequence, another accepted method of classifying the programme priority is one based on cost and speed. In other words, the priority is here defined by what the customer is ready to pay for his requirements. The highest priority will be then given to the customer who will be ready to pay a special price for extra quantum time.

There are many other types of algorithm which may be used in defining the programming priorities, one of them being speed of processing and others storage-space allocation. In Command and Control the priority will be defined according to the importance of the programme and the required speed of the answer.

The normal rule is for the programmes with highest priority to be served first, but of course when a number of programme slices are allocated equal priority rating, they will be handled according to the order of their arrival in the queue.

As explained before, each processing job is allocated a quantum time, which is the maximum time allowed for each programme slice to utilize the computer resources. This time might be interrupted in the presence of a high priority call, which may hold up the process before it has had a chance to run. To avoid this, each programme slice must be allocated a minimum running time before it is allowed to be halted, even in the presence of a high priority. In this way each programme is given an equal chance to be processed until it is completed. Each time it is run, another part is processed, and so on till it is fully processed.

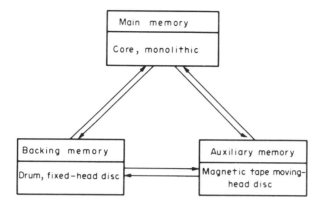

Fig. 9.4. Memory hierarchy in a large computing time-sharing system.

9.4 MEMORY HIERARCHY

Command and Control systems which operate in an on-line real-time environment using time-sharing techniques require very large storage capacity. This is provided by three memory levels, for the systems need not only require a large main memory for the programmes being processed but also large waiting memory units to allow many programmes and large data files to be stored simultaneously. Furthermore, the system must provide fast means of transferring programme slices to the various memory levels. The memory organization for a simple batch processing generally requires only two-level storages, a main and an auxiliary memory. The main memory stores the instructions and the current programme being processed, while the auxiliary memory stores all the relevant data files. In time-sharing systems an extra storage level must be introduced.

In time-sharing systems a large number of programmes may be processed simultaneously. As there is no room in the main memory to build all the concurrent programmes, only one segment (page) of each programme is transferred to the main memory. The rest of the programme bulk, however, must be available for immediate transfer to the main memory upon a demand request. If all the programmes were stored in the auxiliary memory units, large delays would be created while waiting for the transfer. To meet this contingency, a third memory level is introduced which acts as a buffer holding all the current programmes being processed. This level is known as a backing memory. Each time the central processor tackles a different task, a page from one of the current programmes is transferred from the backing memory to the main memory. When the task is completed, the page is returned to the backing memory and a new page from another programme is transferred to the main memory. The size of the backing memory is necessarily very large in order to hold all the necessary programmes.

The various memory hierarchy levels and their interconnections are shown in Fig. 9.4. The main memory keeps the operating programmes, the subroutine programmes which are frequently referred to and the running programme slices (pages). The backing memory holds the bulk of the programmes which are currently operated by the various users. The pages of each programme in the backing memory could be transferred quickly to the main memory for processing, and the full programmes, after they have been processed, may then be transferred to the auxiliary memory for later reference. The auxiliary memory holds all the programmes which are seldom referred to. These include all the historical data files and all the data which does not require real-time handling. In brief, the main memory holds processing data, the backing memory holds real-time data and the auxiliary memory holds all the historical data. If the main memory requires data held in the auxiliary memory for reference, it can obtain it directly without first needing to transfer the data to the backing memory. The memory hierarchy for large time-sharing systems is extended, with each level consisting of a number of modular memory units.

The efficiency of the memory organization is a direct function of the space allocation. When the programme enters the system, the space requirements are not yet known. The best arrangement is to store each page wherever space is available. Thus each programme and its associated data files may be held at various memory locations. Furthermore, each time a page is returned to the backing memory its location may be changed. This scheme can be made operative only if the address of each sliced page is kept in the main memory.

In batch processing, the allocation of memory space is usually according to the programmer's request, but even so the programmer is never satisfied with the space given to him. In time-sharing systems, the allocation of memory space cannot be based on the programmer's request. The whole administrative management in a time-sharing system is different from that of batch processing. In time sharing, the space allocation must be dynamically controlled, as the system continuously shifts items of information to and from the various levels of the memory media.

As already explained, many programmes are simultaneously processed and the total size of all the memory exceeds the physical size of the main memory. The severe problem of locating and handling the various programmes is based on two administrative aids: division into pages and dynamical addressing.

The programme is divided into pages of logical task units and is processed a page at a time, in its correct sequence. This paging arrangement allows small slices of a number of programmes to remain simultaneously in the main memory while the full programme is kept in the backing memory. It ensures the best utilization of the processor time, since the system can transfer at a higher speed from one programme page to another. In this way, transfer can be effected within the main processor without time being wasted waiting for the page to be brought forward from the backing memory.

The dynamic addressing scheme allows each page to be held in a different location in the memory each time it is handled. In other words, it allows reallocation of memory space after each process operation. The main memory holding the operating programme will also consist of special addressing tables which are continuously updated with regards to the exact current location of the pages. Each time a new terminal request enters the system, a new line in the table is opened. When the job is completed, the line is erased and the table is shortened.

With today's computer technology, the memory hierarchy of time-sharing systems has been adopted for most large computer systems. The programmer no longer specifies the physical storage space required, but the system supervisory programme requests the desired space from the memory control. The memory control automatically allocates the space required, and does the calculation for the linking of the needed storage and its address. By the use of a dynamic memory hierarchy allocation, the actual memory available to the programmer may be larger than that provided in the main memory. Indeed, the backing memory may be regarded as an extended main memory. The additional memory space made available in the memory is referred to as a virtual memory. The management of such a memory organization must be based on the following factors:

(a) Dynamic allocation of the data to the various memory hierarchy levels.
(b) Reference tables to indicate the location of the data in the physical memory.
(c) Rapid switching of the pages from the various memory hierarchies to the central processor.
(d) Means of using data from various files by several programmers.
(e) Protecting memory locations from unauthorized access.

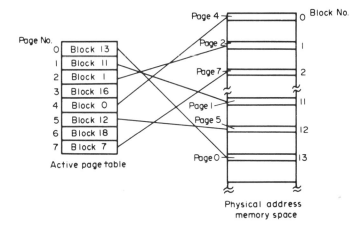

Fig. 9.5. Simple paging organization showing the translation of page address mapping into physical address locations.

The purpose of paging is to make allocation of the physical memory easier. There are many paging techniques but it is only the simplest basic one which is discussed here. In paging, blocks of memory are assigned different page addresses. Through page association a number of non-contiguous blocks of memory can be made to look contiguous and can be dynamically changed by the operating programme during execution. The page table is held in the main memory and consists of one line for each page. Figure 9.5 shows a simple paging arrangement where the data in each line of the table points out the location of the actual page in the memory. The logical address in this example consists of 13 bits, 3 bits for the page number and 10 bits to address one of the 1024 words within the page. These last 10 bits contain the address displacement of the line relative to the beginning of the page. The first 3 bits are used to give access to the page table, which is subsequently translated by means of a decoding matrix into 12 bits indicating the starting address of the currently active page as a block number in the physical memory. The true physical address is then constructed (as shown in Fig. 9.6) by joining the block number with the displacement field.

The page address contains an invalid bit which determines whether the page associated with the page table entry is available. In some applications the page address may also contain the page length.

It can be seen that this paging scheme efficiency utilizes the computer space and saves the programmer's time. The programmer is completely freed of any concern for memory administrative management.

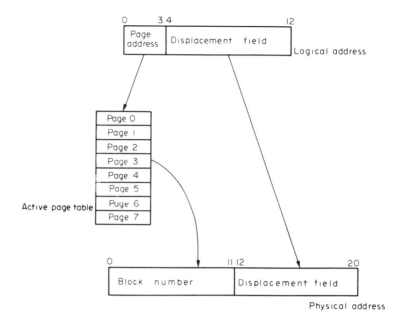

Fig. 9.6. Two-level dynamic addressing schemes.

9.5 MULTI PROCESSING SYSTEM CONFIGURATION

As the computing system grows in size there is an increasing need for a number of programmes to be processed simultaneously. This can be achieved by means of a large computer configuration consisting of a number of processors combined together. This enables several separate but interrelated processing operations to be carried out simultaneously. A computer configuration consisting of two or more processors that can communicate with one another without human intervention is referred to as a multiprocessor configuration.

An example of a multiprocessor configuration is shown in Fig. 9.7, which consists of three central processor units and four independent memory banks. A separate supervisory panel is attached to each processor and enables each unit to operate exclusively. The memory banks are accessible from any of the processors and, furthermore, data can be placed by one processor and read by another. However, reading and writing of data from the same memory bank cannot be performed concurrently. All the data flowing between the units is controlled by suitable instructions from the availability-control unit.

The grounds for preference of a multiprocessing scheme to individual computers or to a large processor are that multiprocessing performs its tasks more efficiently by presenting better availability and better cost performance. It also enables a number of programme tasks to be actively operated in parallel and for the system to communicate simultaneously with a number of remote terminals. The processors may share the load of a single task or may perform a number of tasks and can communicate directly with each other to exchange data. The memory capacity is not restricted to a single processor but is shared between all the processors and is directly accessible from each of them. The last point marks the main difference between multiprocessor and multicomputer

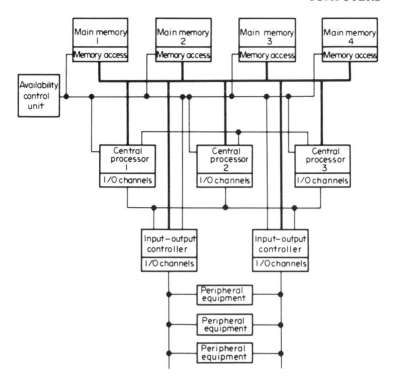

Fig. 9.7. Example of a multi-processing configuration.

configurations. In multicomputers each computer is an individual configuration with its own memory, but while all the computers may be operated together they cannot directly share their memory resources.

A multiprocessor system must also be distinguished from a dual or duplex system where one processor is intended solely as a standby for the other processor in case of equipment failure. Despite the fact that memory could be available to both processors, the same programme tasks are performed in both processors. In multiprocessors the configuration is based on parallel processing which can provide unrestricted multiprogramming of independently written programmes. The collective power of the multiprocessor can bear the work load by increasing system productivity.

As each processor in the system is of equal power and can individually perform all the system tasks, provision is available for dynamically adding or removing central processors or memory banks (as can be seen in Fig. 9.8). The usefulness of this feature is that units may be disconnected for off-line maintenance and the failure of one unit will not disrupt the system operation. Furthermore, the configuration can grow with the increasing development of the system. In the case of failure of one processor the system will continue to perform all the tasks, at the cost only of lengthening the processing time required. (This characteristic of multiprocessing, known as "fail-soft", will be discussed in greater detail in Chapter 14.) The multiprocessor configuration has the advantage of better cost effective operation over the standby duplex processor configuration, since the processing power is always fully utilized. The high reliability of multiprocessing is

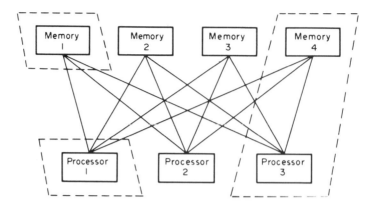

Fig. 9.8. Dynamically multiprocessing configuration (a string of failures or off-line maintenance of a module will not interrupt the active system although will require longer time).

compensated for by the degrading of speed service. Nevertheless, where real-time is the main concern the duplex system configuration is preferable.

The input programmes are distributed to any one of the processors, depending on the application to which it relates. The distribution of the programme is performed dynamically by the availability-control unit.

The multiprocessing potential imposes extra demands on the communication between the various units within the system configuration. The main memory is generally used as a buffer for the data passing between the various processors and between the processors and the input/output devices (as shown in Fig. 9.9). The means provided for sharing the memory between the processors contributes to the efficiency of the system. The designed aim is to increase the number of parallel computation operations. (Since the design of the communication links between the processors and the memory banks is crucial to the system performance, it will be discussed in detail in the following section.)

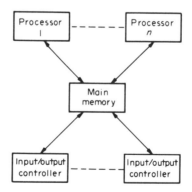

Fig. 9.9. Main memory is used as a communication media between processors and input–output control.

9.6 MODULAR MEMORY ORGANIZATION

In a multiprocessing system any number of individual processor units, memory banks and input/output controllers may be operative in the same configuration. The decisive feature of multiprogramming in a multi-processing system is the ability of the processors to communicate by transmitting information over the shared-memory media. The designed aim of the configuration is to reach a high data-rate transfer between the modules at a low cost and with high reliability.

When two or more processors have access to common memory banks, then data placed in the memory by one processor may be read by another processor. The term processor is used in this section in its general sense to refer to any hardware unit which may communicate with the memory media. That is, it could adequately refer to the central processing unit and any input/output control unit which has direct access to the memory banks.

There are a number of configurations used in various multiprocessing systems for communication between the processors and the memory modules. Each configuration offers a different system serviceability. The consideration of the configuration is determined by performance, speed of transmission, reliability, future expansion possibilities and cost. The proper design of the network architecture of the processor-memory organization is a vital factor for the operational success of the system.

Each memory module can communicate with only one processor at any given time. The communication link between the memory banks and the processor units is formed by a transmission bus consisting of a number of parallel lines for data, address and control signals. At each end of the line there must be hardware equipment to drive the signals along the line; additional hardware is also required to select the necessary bus. The main cost factor of the interconnection network is that of the hardware.

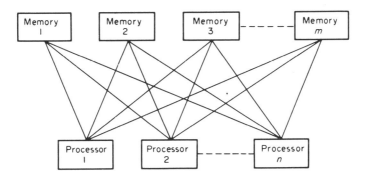

Fig. 9.10. Direct communication between n processors and m memory modules.

The simplest architecture configuration is one with direct lines between each memory bank and each processor, as shown in Fig. 9.10. This configuration is used to a limited extent in simple small multiprocessing systems where it can provide direct access to transfer data between all the modules. This technique is both inefficient and very expensive; in addition, only a very small part of the system may be operative at any given time. The number of buses in this configuration is $m \times n$ but only a maximum of n buses (or m buses, taking whichever is smaller) may be operative simultaneously, while all

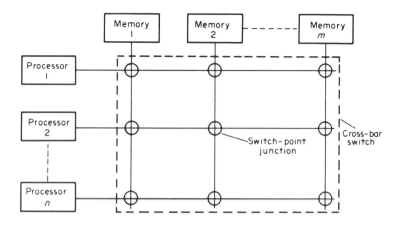

Fig. 9.11. A centralized cross-bar switch allowing communication between n processors and m memory modules.

the other buses are inoperative. Furthermore, expensive complex hardware is required at the access to each module, to select the desired bus. For this purpose two types of hardware circuits are available, one consisting of a set of line drivers for each bus and the other a set of line drivers which can be directed to the desired bus by a logic selection matrix. A further drawback of this configuration is that it is a static network, as each time an extra module is added the hardware selection circuits must be modified. It is obvious that this configuration is unsatisfactory and that other techniques should be used.

A second architecture approach is to centralize all the hardware communication switching. This configuration is based on the cross-bar switch, as shown in Fig. 9.11. The cross-bar arrangement seems to be a much better technique than the direct technique, as it may provide better efficiency and be basically cheaper. It requires only one set of drivers at the access to each module, thereby rendering the system simpler than with the direct technique. In order to receive a connection, a switchpoint at the junction of two buses is operated. The cross-point consists of $n \times m$ switchpoints, which are the number of junctions between any two perpendicular buses. The whole cross-bar operation is controlled by a control unit which may allow up to m (or n) junctions to be switched simultaneously.

Despite the superficial advantages of the cross-bar switch, it suffers from severe drawbacks in its practical applications. The main disadvantage of this technique is in its prime characteristics, that is, in its highly centralized concept with the cross-bar performing the crucial link within the system and thereby drastically reducing the reliability of the whole configuration. A failure in the main cross-bar switch will cause the entire system to fail. Duplication of the cross-bar equipment would ease the reliability problem but would not solve it, and moreover would double the cost to a point which would make it more expensive than the direct method. Furthermore, since a cross-bar switch may be reached by a number of modules simultaneously, different data-transfer requests may result in conflicting switching requirements. To solve this contingency, a priority selection scheme would have to be introduced, but this would complicate the common control switch. Another disadvantage is that the cross-bar switch, similar to the

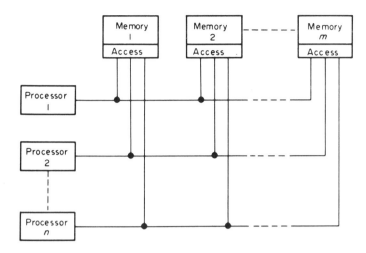

Fig. 9.12. A distributed cross-switch organization for connecting the processors and memory modules.

direct technique, suffers from a rigid construction. The architectural configuration cannot be allowed to expand beyond that which was originally designed.

A third architectural configuration is based on a combination of both previous techniques. The principle of logic behind the cross-bar switch is retained, although the common logic circuitry is distributed between all the memory modules. The resulting configuration is shown in Fig. 9.12. It is a more reliable scheme, as it is based on distributed logic which allows a number of failures to occur without greatly affecting system performance. It is a flexible combination configuration because further memory banks may be added without affecting the wiring of the other units. A failure of a memory bank or a processor will still allow the remaining system modules to carry on with the work load. Special attention is given to the availability consideration. Should an access request conflict occur between the processors, the memory-access unit grants storage access to the processor which has the highest priority. This configuration is most popularly used, particularly in large multi-processing systems, because it improves the factors of cost, reliability and flexibility.

System reliability is increased here, as the memory-access unit is independent of the memory modules involved. The bus connection to the memory-access could also be made to be independent for each bus by using separate ports. In the latter configuration, the front end of each memory module consists of a memory access unit which comprises n ports, one for each processor, as shown in Fig. 9.13. The scheme for selecting the ports may be implemented by multiplex or multiple simplex techniques. This entails having the buses from each processor chained through all the memory access units in turn. This arrangement allows system flexibility, as the expansion only requires the addition of ports. It also allows certain processors to communicate with particular memory modules.

The main disadvantage of the distribution communication technique is in the massive wiring needed to implement the scheme, resulting in propagation delays which consequently reduce the access and cycle time of the memory. Means must therefore be introduced to reduce the number of buses and consequently to call for less cabling.

It was pointed out earlier in this section that the term processor could also refer to

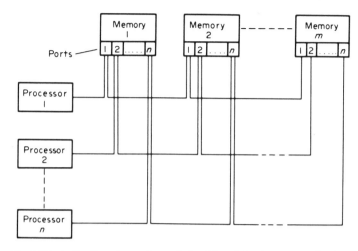

Fig. 9.13. A distributed cross-bar switch with a chained-port organization.

input/output modules. As, however, the data-transfer rate from the input/output modules is far slower than that of the processor and the memory access bus interface of the distribution communication configuration is standard for all the channels and control ports independently of the modules involved, this makes the bus connecting the input/output modules extremely inefficient. To increase the efficiency of the system the transfer rate of all the buses must therefore be brought to operate at an equal speed. This is achieved by having a number of low-speed modules sharing the same bus by means of a centralized multiplex switch. Figure 9.14 illustrates the arrangement where a number of input/output modules are clustered so that all the buses in the distributed network have an equal transfer rate. To be effective, all the modules connected to the same multi-bus must operate at the same speed. In each shared connection there is a multiplex access control unit which allocates the multi-bus according to an internal priority. The maximum efficiency is obtained when the volume of data flowing in the bus is near or just below the saturation figure.

A fourth configuration approach is based on the loop network described in the previous chapter. The loop operates by using a high-speed multiplex, where a serial stream

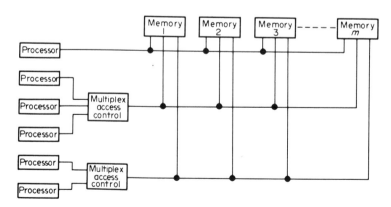

Fig. 9.14. A distribution cross-bar switch with a shared-bus technique.

of bits flows in the loop. A number of bits, headed by a sync. bit, represents a frame, i.e. a time slot. The operation of the loop is controlled by a loop synchronizer which provides the basic timing for the loop. The synchronizer modifies all the timing pulse shaping without altering the information contents in the frames. All processors and memory modules operating at an equal high speed are connected in series in a single-loop configuration, as shown in Fig. 9.15. Slower-speed modules employ different loops which are connected to the main loop by means of loop couplers. The speed at the secondary loops may operate at lower rates than the main loop. All the modules have a unique address which follows the synchronous bit at the head of the time slot.

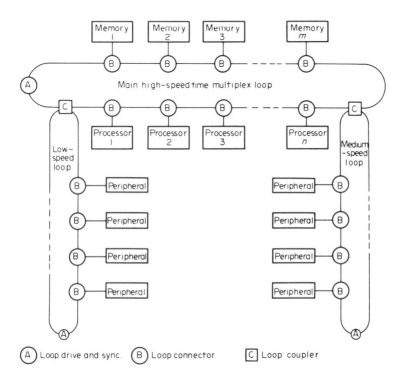

Fig. 9.15. A loop time-multiplex organization used to connect processors and memory modules.

9.7 HIERARCHY OF MODULAR COMPUTER CONFIGURATION

The principle of multiprocessing requires that the network configuration contain a number of equally powered processors to share the system task load which, in effect, means that each processor may perform all the tasks, although then it will require longer processing time. Where the system is intended to execute more than one programme, the basic method to achieve more computing power is simply to couple a number of processors to share a primary memory. In designing a multiprocessing system, its size should be gauged according to the largest task it has to perform. This does not apply to all applications, as in many cases, if the modules are not geared to the tasks, huge overhead costs would be later incurred.

Basically the aim of the design is to produce a system suitable for application to specific tasks while preserving the economic benefits of utilizing standard units. This calls for a modular computer system where each module performs specialized functions, thus effecting an improved overall cost–performance ratio.

With the increase in size of Command and Control systems, the networks tend to become more and more sophisticated. More multiprocessors are needed to tackle large tasks, with each processor performing a portion of the same task; in Command and Control systems, dedicated processors are required where each processor controls a separate type of task. The best example here is the complex communication network. As it would be wasteful to give this task to a standard processor, a special input/output processor is used. Considerable saving is achieved by tailoring the processors to meet the specific requirement of each level, for each processor is relieved from the routine tasks which can be performed by the other processors.

Many modular computer configurations are to be found, varying from a multi-special-processor systems to a multicomputer systems. There are systems where all the computing power is concentrated in one location, while in others the computers may be miles apart. In all the configurations there must be means whereby all the processors can communicate with each other. The various types in use will be dealt with in the following section.

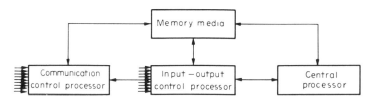

Fig. 9.16. Hierarchy of processing power.

A simple dedicated three-processor system is shown in Fig. 9.16. Each processor has its own arithmetic unit, although they all share the same memory media. The central processor handles all the computing and updating functions, such as file management, programme compilations and programme execution. The input/output processor controls the traffic to and from the central processor, thus releasing it from all the routine tasks. The 1/0 processor, known as "front-end" processor, presents the data in computer form; in other words, it translates codes, edits the data and builds up the blocks. This processor can also control all the local peripheral equipment, including all the auxiliary memories and all the local terminals. The third processor is required when there are many remote intelligent terminals. This processor has the duties of message switching and controlling the communication with all the remote links by performing polling and selecting operations. The communication processor can operate independently, regarding the other two processors as hosts.

The basic approach shown in Fig. 9.16 is a well-established technique as a means of dividing tasks between an input/output processor and a central processor. Each processor is designed to perform its particular task independently, although the results are intended to assist the other processors.

The approach where the tasks are shared between a number of processors could be extended to where each task is allocated a different computer. In this type of configuration each processor operates as a separate enterprise, although eventually their computational results are transferred to the other computers in the network. The distribution of computing power is contrary to the accepted view where a computer must be defined according to the largest programme which is to be run. For a time-sharing system with variable-length programmes, the conventional approach could be most uneconomical, as the expensive system equipment, including all the peripherals, is divided between all the customers, even those whose requirements are unappreciable. In the other approach the big and long programmes could be handled by a special large computer, while all the other small programmes could easily be handled by a minicomputer. This new approach not only efficiently distributes the computer power but also increases the system reliability. The design of these systems is emphasized by the modularity of the tasks on their implementation. The Command and Control system engineer should be on the alert when approached by the computer salesman with offers of large computer configurations which can perform all his tasks. However genuine the offer, the result is not always economical. A series of computers of variable sizes and configurations, where each computer performs an independent task, is more practical. This division of computer power between a hierarchy of computers is a very important factor for the design of Command and Control systems. An example of a multi-computer configuration used in a Command and Controlled system is shown in Fig. 9.17. As the tasks are not completely

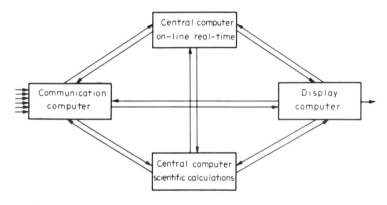

Fig. 9.17. Hierarchy of computing power.

identical, each module may need a particular type of computer with different computing power, processing speed and memory capacity. In the referred to example, one of the computers tackles all the real-time on-line tasks and transfers the results to another computer, which performs all the off-line analysis calculations. The scientific analysis generally requires long processing operations, which are better performed by a separate installation, as they would only interrupt the real-time operations. A large-screen display requires a separate computer, since its tasks are completely different from those of the other computers in the configuration. This computer performs negligible calculations, as its main task is to control the display equipment, leaving the new data to be fed by any of the other computers. The communication computer is an independent message-switching module which regards the other computers in the same way as it does a remote centre.

A variation of a multicomputer configuration is the duplex (or triplex) system where

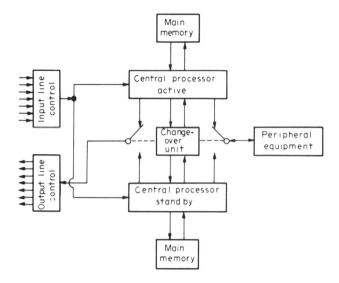

Fig. 9.18. Duplex computer configuration.

two computers work in parallel, although this is dictated by reliability requirements rather than by operational requirements. In this configuration both halves of the configuration bear the full task load, with one-half active while the other is on standby, as shown in Fig. 9.18. In multiprocessing, all the processors share the task load, and with a failure of one processor the other processors can continue to provide the services with but a slight lengthening of the operation time. The reliability factor in multiprocessing is quite adequate for most operations, except where real-time is involved. For real-time applications a duplex configuration is essential, with the standby computer fully operative, ready to take over with negligible delays when the active computer fails. The main memory too must be separate for both halves of the duplex and so cannot employ the memory techniques which were described for multiprocessing. The reason for this is the need to ensure that the real-time files will not be updated by two programmes at the same time. The real-time system calls for reliable parallel data base files in each half of the system.

The hierarchy of computing power could be extended to a distributed computing

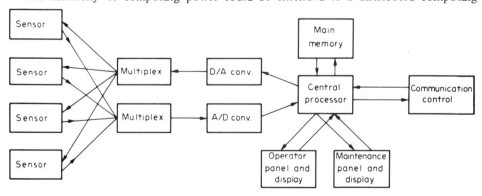

Fig. 9.19. Remote telemetry computer installation.

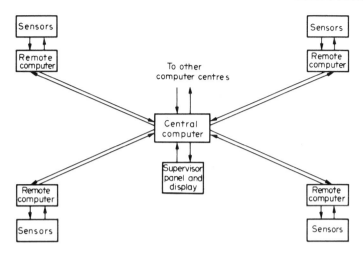

Fig. 9.20. Distributed computer power.

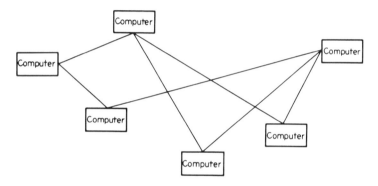

Fig. 9.21. Distributed computer resources.

power where small computer installations are located at remote centres. This scheme is very popular in telemetry schemes where in each remote location there are a number of measuring equipments operating in a separate configuration. A local remote computer controls all the telemetry equipment (as shown in Fig. 9.19) consisting of data acquisition, data reduction and process monitoring. That is, it collects the measurement data, compares it with previous readings obtained from the same location and also commands the sensors according to a local programme or to centre instructions. The local computer transmits to the centre computer only new relevant data or ambiguity measurements. The local remote centre may operate automatically, i.e. unmanned, although provision is provided so that it could be controlled locally. One application for such a scheme (see Ref. 20) is an air-traffic control where each local centre controls a radar system. The information collected in the remote centres is transmitted to a centre location (as shown in Fig. 9.20), where it is correlated with data collected from other remote locations. The new data is used to update the central file and to build a status real-time picture of the air-traffic flow. The central location may be part of a larger system of configuration. That is, each centre may be a regional air-traffic control with means of communication with other regions to coordinate the flight of the planes as they move from region to region.

As the Command and Control system grows in size, there is a need for the distribution of computer resources when computer centres are located at widely spread geographical locations (as shown in Fig. 9.21). Each computer operates as a completely independent unit, although there is still a need for possible communication between them. A practical example of this configuration could be the regional air-traffic control discussed above. This type of computer configuration is finding many new applications. It enables each centre to perform local processing and also enables all the computers to share the resources of other computers. An important aspect of this resource distribution configuration is the reliability of the system. The structure has safety potentiality which is not found in a centralized system. Owing to the important aspect of this type of multicomputer configuration, it will be discussed in the following chapter in greater detail.

9.8 REFERENCES

1. Watson, R., *Time Sharing System Design Concepts*, McGraw-Hill Book Co., 1970.
2. Head, R.V., *Real Time Business Systems*, Holt, Rinehart & Winston, Inc., 1964.
3. Black, W.W., *An Introduction to On-line Computers*, Gordon & Breach Science Publishers, 1971.
4. Desmonde, W.H., *Real-time Data Processing Systems: Introductory Concepts*, Prentice-Hall, Inc., 1964.
5. Riley, W.B., "Time sharing: one machine serving many masters", *Electronics* (29 November, 1965), pp. 72-78.
6. Sackman, H., "Time-sharing versus batch processing: the experimental evidence", *1968, Spring Joint Computer Conference, AFIPS*, Vol. 32, pp. 1-10.
7. de Mercado, J., "Introduction to time shared systems", *1972, International Seminar on Remote Data Processing, ILTAM Israel*, Pt. 1, pp. 159-181, 279-314.
8. Dudick, A.L. and Pack, C.D., "Round robin scheduling in a computer communication system", *ACM/IEEE Second Symposium on Problems in the Optimization of Data Comm. Systems*, 1971.
9. Coffman, E.G. and Kleinrock, L., "Computer scheduling methods and their countermeasures", *1968, Spring Joint Computer Conf., AFIPS*, Vol. 32, pp. 11-12.
10. Arndt, F.R. and Olivier, G.M., "Hardware monitoring of real-time computer system performance", *Computer*, Vol. 5, No. 4 (July/Aug. 1972), pp. 25-29.
11. Westerhouse, R.A., "Time sharing and its applications", *Computer*, Vol. 2, No. 7 (Jan. 1969), pp. 3-7.
12. Flynn, M. and Podrin, A., "Shared resource multiprocessing", *Computer*, Vol. 5, No. 2 (Mar./Apr. 1972), pp. 20-28.
13. Laliotis, T.A., "Main memory technology", *Computer*, Vol. 6, No. 8 (Sept. 1973), pp. 21-27.
14. Dennis, J.B., "Segmentation and the design of multiprogrammed computer systems", *1965, IEEE International Convention Record*, Vol. 13, Pt. 3, pp. 214-225.
15. Corbato, F.J. and Saltzer, J.H., "Multics – the first seven years", *AFIPS, Spring Joint Computers Conference*, Vol. 40, 1972.
16. UNIVAC 1108, "Multiprocessor system", 1968, Sperry Rand, UP-4046, Rev. 2.
17. Blàauw, G.A., "IBM system/360 multisystem organisation", *IEEE International Convention Record*, Vol. 13, Pt. 3 (1965), pp. 226-235.
18. Ellis, R.A., "Modular computer systems", *Computer*, Vol. 6, No. 10, pp. 13.
19. "C. System general description", Collins Radio Company, 523-0561697-20173R, May 1970.
20. Ben Hiat, R., "The E-A 1000 multi-level computer control system", *IEEE International Convention Record*, Vol. 13, Pt. 3 (1965), pp. 176-187.

CHAPTER 10

Distributed Computer Resources

10.1 INTRODUCTION

Distribution of computer power in a single network is based on sharing the resources of the computers in the configuration, where each set of computers may be either independent systems or dependent subsystems. Each computer installation may be adjacent to the other computers or may be miles apart with the whole network spread over wide geographical areas. Nevertheless, they all have the characteristics of being able to share the resources entailed in the other computer installations as if they were part of the same computer installation.

The area of resource sharing, although a relatively fresh field, is rapidly growing, and there are indications of considerable expansion in the future. The sharing of computer resources is possibly the largest contribution to the advancement of science, for it permits the sharing of both the programming facilities and the scientific results, thereby enabling the processing of large complicated calculations by drafting to the task all the computing power in the network. This is something which was impossible with a single installation, irrespective of its magnitude.

A resource-sharing network is essentially intended for Command and Control applications which have their resources dispersed over wide geographical areas, but it is also designed to provide extra computing power to those locations which lack the backing of large installations. There are two basic types of applications which may require the sharing of computer resources. In the first application, there is the need of sharing the information of files of different independent or dependent installations. In the second application there is the need of sharing the computer power, where each installation can perform and operate its local programming problems, while all the installations may be operated together on solving more complex programming problems. The large and long programmes could be handled either by a special computer in the configuration or by joining a number of computers together. All the other small programmes could easily be handled by local computers.

In a single computer installation, the computer power required is defined by the largest programme to be run in the system. This is a most uneconomical arrangement in time-sharing systems, as the whole system equipment cost including the peripherals is shared between all the users, even by those who do not require the use of all the equipment. In addition, these expensive peripheral equipments may not always be fully exploited. In distributed shared resources configurations, each computer location is defined by only the average small programme, while the larger programmes could be run by joining a number of computers to share resources. In such a configuration the peripheral equipment may be situated in only a few of the centres but with access facilities from all the centres.

The facilities which are presented by the shared resources are as follows:

(a) Presenting a wider range of system resources to a larger community of customers.

(b) The gathering of specialized demands in a particular designed centre and the pooling of the resources between the centres.

(c) Sharing major expensive resources across organizational boundaries.

(d) Extending local files to include extra data available at other centres.

(e) Transfer of data to customers associated with other centres.

(f) The ability to present to each customer access to local, regional and national information-storage files.

(g) Distribution of intelligence resources.

(h) Distribution of equipment which consequently reduces the cost of the local programming.

(i) Provides communication with larger computing systems in order to perform complicated compilations which are impossible on the local computer.

(j) Provides means of coordination with other computers, to share the efforts in solving complicated computing problems.

(k) Allows exploitation of peripheral equipment, computing resources, file resources, and information-handling equipment which are situated at other centres.

(l) Builds up large programming libraries accessible to all users. Each local centre retains all the programmes which are vital to its operation, together with lists of programmes held at other centres.

(m) Builds up unique software programmes available for all users.

Administrative message-switching network is a popular example of a distributed computer network, although it is not a shared resources configuration. Nevertheless, message-switching centres are used as tools to achieve resource sharing. That is, the communication computers used in the configuration act as message-switching agents which transfer the shared data for distribution among the host computers of the system configuration.

The best configuration facilities of resource sharing are presented by a diverse set of computer types rather than a homogeneous set of computers. By using different computers in the same distributed configuration, each local centre could deal with a much wider scope of problems than could be possible in a homogeneous configuration. The design of each host computer in the local centre should be according to the local processing requirement and not according to the largest programme to be run in the overall system. Nevertheless, the system must enable large programmes to be run either by one specially large computer or by distributing the load over a number of small computers.

The complexity of interleaving independent systems is enormous and imposes a considerable burden on the integration design. In theory, the system design could regard the interconnection links with other systems as another input or output, but this approach is complicated by the fact that each system may use different computers operating at different speeds and programmed by different languages and procedures. The problem could be defined as the handshaking requirements between different computer system centres, where each system may be operated on completely different levels with little in common with the other systems apart from the fact that it contains information essential for the operation of the other system. The only way to overcome the handshaking problems is to use special dedicated communication computers as interfaces between the various centres, as shown in Fig. 10.1. These interface communication computers act as message-switching units, while the data processing is performed by host computers which are independent system centres.

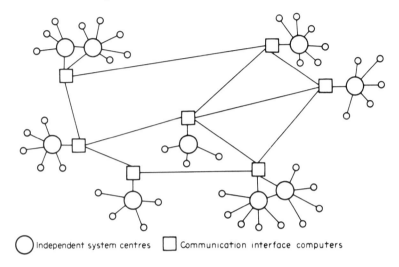

Fig. 10.1. Distributed system network with shared resources.

10.2 THE IMPACT OF SHARED RESOURCES

With the adoption of shared computer resources many new problems arise and call for consideration. Not all of them are confined to the engineering fields, for many encroach on other areas, such as the social and the political spheres. The first problem to be tackled is how to integrate the various distributed computer installations into one configuration. Only when this and all the other initial problems have been solved can the full possible impact of the introduction of shared resources be revealed.

In a distributed shared resources configuration, each computer centre generally operates independently and is responsible to its own file management. The main questions to be then solved are who is responsible for sharing the resources and how it can be performed. If one centre wants information from another centre, who grants it the facility? Are all the files accessible to all the customers, and how can security of the data be adopted and preserved? If data is collected by many terminals at various centres, who will be responsible for updating all the associated files? There is no single answer to all the questions, for it generally depends on the specific application the network is used for. Nevertheless, some design suggestions are presented here.

Each information file can have only one controlling agent, so the files can be updated and amended by only a single hand. In other words, each computer will be responsible for its own files and only it can grant access to them. In some system, such as the air-traffic control described in the previous chapter (see Fig. 9.20), there may be a number of computers operating together in a local centralized system. In that case, one computer could update the files and change the instructions of other computers in the network, and then the first computer would be termed the master and the other computers would be referred to as slave computers. A slave computer might be associated with only one master computer, which nevertheless may be in association with a number of slave computers. The slave computers perform the data collection for the master computer but the updating of the latter's file is its own responsibility, which it performs only after correlating the data received from various sources. On the other hand, the master can change all the slave files, and its transmitted instructions must be regarded as commands. These design suggestions are intended to prevent a particular computer from being presented with conflicting instructions.

In a distributed shared-resources configuration there may be a number of master computers which are all operating independently although they have means of communicating with each other. While each computer may share the information files of another independent computer, it cannot cause any updating or any changes in these files. Should an independent computer require information from another, it can receive it only if the other computer is ready to grant it. If it should require continuous data from another computer, this operation could be inserted into both systems as a permanent task. The receiving computer will then regard this input as if it arrived from one of its data-collecting terminals. That is, the channel of data transfer will operate in a simpler form, enabling the files to be updated in the same way as from the other data-collecting terminals. However, the fact that a computer gets continuous data from another does not give the receiving computer the right to cancel or change the data files of the transmitting computer; nor will the receiving computer have the right to check the operation of the other computer or check the authenticity of the data presented to it.

Attention must be drawn to the fact that sharing resources causes further types of problems which are more social than technical. The main problem is the interpretation of the same data, which may have different meanings at the various locations. A simple example is that of time, which is a term liable to misinterpretation because for each computer its reference is always to local time, and in a widely distributed computer network there would thus be reference to a number of distinctive times. In consequence, the variation necessitates the conversion of the time element each time the relevant data is transferred to another location. In many Command and Control systems one centre may require data from many different locations each having a different time reference. Furthermore, in a "real-time" distributed environment this could raise many design problems. Before tackling this problem, an answer must be given to what "real-time" means, as the establishment of this point will have an effect on the whole operation. It is suggested that any translation of time (or any other similar items) should be performed where the data is going to be used and not where the data originated from. The local files must therefore always be kept according to the interpretation which is given by the local base. This requirement will have the effect of building up and keeping special tables at each centre which will define the translation rules of all the items according to their source. Another solution to the general interpretation problem could be the enforced adoption of standard technical meanings for all the terms used throughout the system configuration. This in effect would require all the files to be edited, with the relevant items translated before they could be used locally. For example, 'time' will always be referred to as B.S.T. in all the files and then each centre will interpret it according to its local reference. It is obvious that whatever the solution, the translation requirement may have quite an impact on the local system design. Here only the example of time was discussed, but many other items could be shown that have similar effects.

Another problem to be solved in a shared resources configuration is how to combine computers so that they can share in the execution of the same task. At first glance the solution to this problem would appear to be similar to the way it is solved in a multiprocessing system. This consideration, however, in a distributed configuration is not the same as in a multiprocessing system, where special hardware is inserted to serve as the availability control unit (see Fig. 9.7). The question here is who decides what part of the task each computer will process at any given time and who controls the task sequence operation. To solve this, one of the computers in the configuration will have to act as the availability control unit and thus be responsible for implementing the task.

The development of shared-resources calls for much closer cooperation between the various organizations and closer liaison between the different countries associated in the system. This demands a much greater need for standardization and control disciplines to make possible better quality and services. These demands will range from the use of standard language procedures to communication network protocols. There is also a need to define special languages for various system operations which may be operative on the different computers available in the system. These standards must define strict procedures for gaining access to each file and locating the items in the various page maps. Other standards must be adopted to create a uniform scale for timing, pricing and similar policies.

One of the most serious problems that could occur in a shared resources configuration derives from its main advantage. There is a possibility that many independent systems will in practice find themselves unduly dependent on each other. As long as each system was operating independently, it was its own responsibility to check and verify all the information items it uses. Once a number of independent systems join together they will become dependent on each other's resources. They will now have no control of the resources and will have to accept them in the form the other system is ready to present them. The local system centres, which were once completely independent and still capable of operating on their own, will gradually become dependent on the other systems' facilities. That is, they will get the information without any facilities for checking its accuracy. This drawback stems from the fact that each system which supplies information to others will still want to maintain its privacy and prevent any infringing of its information sources and procedures. These disadvantages could cause the individual system to lose its local responsibility and this could consequently cause reduction in the efficiency and in the whole local system effectiveness. If task faults arise in the shared system, the question will always be raised, who in fact is responsible?

To conclude, it is clear that distribution control can only be effective if there is some adequate central control. The control need not be restricted to technical matters but must cover also operational procedures, including standardization of protocols.

10.3 COMMUNICATION INTERFACE HIERARCHY

The full exploitation of the potential power of resource-sharing systems can only be achieved by merging both computers and communicative networks in a single configuration. The issue is not a matter of relating these two mediums but of blending them into a single sophisticated system. The computers are then no longer nodes in the communication network but the agents managing the communication processes. These master functions may either be conducted by the destination computers or may require specially constructed computers. Before discussing the communication management in a resource-sharing configuration it is essential first to discuss the hierarchy of communication processing.

The connection between computers with remote terminals via communication lines has already been discussed in detail, in the previous chapters. There are three distinct levels in obtaining the communication connection, as shown in Fig. 10.2 (b).

(a) Circuit switching, which provides the physical connection.
(b) Logical path connection, which is provided by polling the physical path to select the required terminal.
(c) Information transfer, which is provided by a computer.

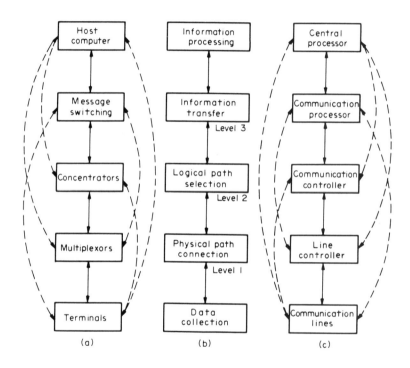

Fig. 10.2. Communication interface hierarchy.

On the lower level the circuit switching is performed by multiplexors which select the physical path, i.e. the lines, each of which may have a number of terminals attached to it. This level is usually located near the terminals. The second level is normally a remote centre which requires more sophisticated equipment, generally provided by a minicomputer. This minicomputer performs all the concentrating line duties for selecting the logical transmission path. The actual handling of the information is performed on the third level, and calls for specially orientated computers. This third level, which is the central site, is provided for by message switching. It should be noted that these three levels do not necessarily work together or in the sequence stated, but may operate singly or even in a different order of combination, as shown by the dotted arrows in Fig. 10.2 (a).

The data collected at the terminals is generally intended to reach the destination of the host computer so that it can be processed. The information received at the central processor must pass all the three level transformations before it may be used. This means that demultiplexing and/or deconcentrating duties must be performed at the computer end. The information must be first brought into a form which permits it to be handled by the central processor. The input serial data must be accepted in its proper sequence and the associated address recognized before an interrupt action can take place. When a message is transmitted from the computer to the terminal, the relevant line must be selected, the information built into the proper format and the correct timing inserted. These tasks may be performed by the central processor or may be transferred to a separate processor adopted for the purpose of handling the communication tasks.

The communication management within the central processor is not identical with the one discussed before, although there are many similarities. The three levels in the communication processing, as shown in Fig. 10.2 (c), are intended for dealing with the efficient handling of the vast number of processor input/output lines. By locating these extra levels outside the central processor, the central processor may be freed of most of its communication tasks and thus permit the data gathering to be isolated from the data processing. It must be pointed out that there are many forms and combinations of communication processing, but only the basic ones are discussed here.

The lowest level in the communication hierarchy is the line controller (shown in Fig. 10.3), which is the interface unit between the central processor and the various

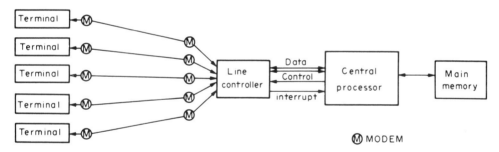

Fig. 10.3. Line controller interface.

communication lines. Similar to the lower level already discussed, the line controller is essentially a multiplexor which performs the physical selection of the lines. The line controller, however, has many more duties connected with its interfacing tasks, the equipment of which can be provided either by hardware and/or software. The main hardware difference is the introduction of control circuits and buffer storage registers of character size for each output line.

The line controller provides the modems with all the required control signals which are needed both for communication lines and the modem operation, and prepares the data received from the lines to be fed into the CPU. It converts the bit serial data received from the line into parallel characters to be fed into the CPU, and performs the speed conversion. To execute these operations it must have enough logic to recognize the start or synchronous codes and any other communication codes. After the full character has been received in the buffer, the line controller sends the CPU an interrupt flag signal, and when permission is granted it transmits the relevant character with its associated address. When the computer sends out a message to the terminal, the line controller transmits it in serial form, bit by bit, a character at a time. The line controller must first select the required line and then organize the message in the right timing sequence, adding all the relevant communication protocol signals.

The line controller only handles one character at a time, while all the main communication tasks are still performed by the central processor. Furthermore, the central processor must clear the buffer registers in the line controller before another character can arrive. This limits the usefulness of the line controller to only a small volume of traffic.

In order to cope with a larger volume of traffic and further reduce the load on the central processor, a separate processor is used for controlling the communication (as shown in Fig. 10.4). This communication processor, which is also known as input/output

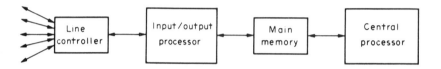

Fig. 10.4. Communication controller interface.

processor, performs additional communication functions and has the facility to initiate transmission procedures such as polling and selection. In contrast to those of the line controller, the functions of communication controller are accomplished mostly in software. This is practical, because it has a much larger buffer storage and so can hold a full block of characters. The serial bit data, which arrives a character at a time in a random order, is built into a block of associated pieces of information so as to transmit them together to the next stage. The function of block assembly, which was previously th᾽ task of the central processor, has been shifted to the communication controller. In this way the assembled block can be placed directly into the main memory with no need for the central processor's intervention. Before the block is transferred to the main memory, the communication controller performs error detection and/or correction of the received data. If an error is detected it may either correct it or may require retransmission of the block. The communication controller also performs basic editing functions such as blank deleting and code conversion. In the transmission mode, the information is brought into the correct format, and the required checking codes and all the necessary communication protocol characters are added.

The communication controller is a very primitive processor which only acts as an interface to assist the central processor, and leaves all the actual message handling and any further communication processing tasks to the central processor.

With the increase in the traffic volume, better efficiency can be reached by completely isolating the communication processing from the central processing. This is achieved by a stand-alone computer which has as its prime task the communication processing (as

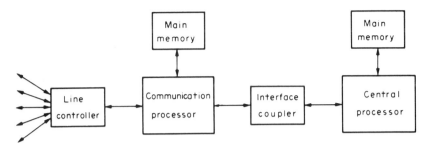

Fig. 10.5. Communication front-end processing.

shown in Fig. 10.5). The central processor is now completely relieved of any communication tasks and can devote all its resources to information processing. The communication processor, also known as front-end processor, is responsible for all the handling of complete messages, and thus is basically a message switching unit. The front-end processor provides efficient operation and the required throughput to the central processor by off-loading from it all communication-type functions, thus freeing it for purely computer processing tasks. The two processors can communicate with each other via an interface coupler which can initiate high-speed parallel data transfer so as to support and fully exploit both systems. This configuration is a form of parallel processing which is a much cheaper system than with a multiprocessing capability. The front-end processor can handle all the incoming messages, performing both priority and queueing functions which were previously performed by the central processor. It can also perform basic processing, such as communication statistics, selection of alternative routing, and message heading analysis, as already discussed in Chapter 6 for message switching.

10.4 COMPUTER TO COMPUTER COMMUNICATION

The essential purpose of resource sharing is to facilitate communication by enabling each computer to communicate with any other computer in the same configuration. Beyond this, it generally possesses decentralization characteristics where the configuration is always dynamically expandable, thereby allowing the addition of extra computers of various sizes and different computing power located far apart.

Computer-to-computer communication requires a different design-approach from computer to terminal communication. The transmission channels require the transfer of large bulk files, and consequently much faster data transfer rates are called for. Transfer rate of 50 K bits/s are generally used, although faster systems are also available. The required fast channels need to be switchable, a facility which can only be provided by special communication processors.

A further reason for the need of a separate processor is that the communication tasks in computer to computer connection involve many overhead processing operations. This point is not a question of system cost but of the efficient operation of the configuration. If the communication tasks are left to the central processor, little time will be available for it to perform its own processing task. The design aim should therefore be directed to relieving the central processor of most if not all the processing tasks. The main processor would then have to deal only with the actual information and will not be concerned at all with priorities or with the arrival time of the messages. It is for the communication processor to assemble the messages, and only when the data is completely assembled will it pass it to the larger central processor for the required process of manipulation.

The communication processor is responsible for all the communication management of transferring the large messages between the sender and the receiver. It performs on-line testing of all the lines and the task of informing all the other communication processors in the configuration of the network status. The communication processor keeps a statistical record of all the traffic which flows through, including error rates, number of failures, etc.

There are three basic types of communication processors:

(a) Front-end processors, which are the interface between the host computer and the communication lines.

(b) Stand-alone processor, which is an independent message-switching centre handling the communication traffic from a number of separate computers.

(c) Remote communication processors, which perform the main tasks of linking a number of communication links.

The main advantages of communication processing may now be summarized as follows:

(a) Increasing the efficiency of the traffic transmission on the communication lines.

(b) Releasing the associated host computer from any communication tasks.

(c) Reduction in the required processing time.

(d) More economical for the management of large communication configurations.

(e) System flexibility and modularity, which allow it to expand by adding more centres and more host computers without requiring the change of any hardware circuits.

(f) Reliable operation, which prevents the failure of the communication processor to affect the host or configuration operation.

(g) Enables the off-line maintenance of the communication processor without affecting the host computer operation.

(h) Very high availability, incorporating redundant systems.

(i) Controls different type of lines, speeds, codes and terminals without the assistance of the host computer.

(j) Keeps on-line checks of all the lines and all the terminals to ensure proper operation.

(k) Keeps statistical records of all the communication transactions.

The communication processors are placed at the various nodes of the network topology, to allow resource sharing of their host computer. There are three alternative network topologies control which are used in resource sharing:

(a) Fully distributed control, where each centre operates independently although all the centres may be either fully or partially connected.

(b) Centralized control where all the distributed centres are controlled by a central processor supervision.

(c) Loop control, where all the distributed centres are connected in a ring. This scheme is a mixture of both centralized and distributed techniques.

The basic difference between the communication network discussed in the previous chapters and the communication network which consists of communication processors is in the design approach. Instead of designing the communication system around the computers, the design approach here should be directed at regarding the computers as one of the components of the communication system. Computers should be inserted into the network wherever they contribute most to the efficient operation of the system.

10.5 SYSTEM NETWORKS

Today there are many computer-shared resources in operation or being developed, both commercially and experimentally. They are being increasingly required to meet the unanticipated growth in traffic volume between computer installations. In one of the systems, for instance, a further growth of up to 26% per month was recorded over a period of 3 years. (This is a challenging situation for the technology of the subject, which may be said to be still in its early stages of development.)

Of the many systems currently in operation only three of them are presented here to indicate the nature and range of the new expanding field.

ARPA-NET

This network has been termed ARPA-NET after its sponsor, Advanced Research Projects Agency Network. Theirs is the most advanced work in the field of distributed network, designed both to explore network technology and to interconnect research centres for the sharing of their resources. The ARPA-NET allows intercommunication between dissimilar computers at widely separated installations. It is intended to permit scientists and programmers at a research centre both to have access to data files and to use interactive programmes which exist in other centres. The ARPA-NET is a nationwide system which is designed to allow the use at one computing facility of particular hardware and software facilities for communicating with another remote computing facility. The system uses 50 K bits/s channels to interconnect the computers at the various universities and research institutes throughout the U.S.A. Today it encompasses also other countries, such as Norway and Britain.

The ARPA network is a distributed configuration where the node interface is constructed by communication processors called IMP (abbreviated from Interface Message Processor). The IMP is intended to service the local host computer and act as an intermediate communication interface transferring messages to their specified destination. Initially each IMP was destined for, and required to service only one host computer, although means are provided for a single IMP to service up to four host computers. The basic ARPA network is shown in Fig. 10.6, which illustrates the way the host computers are interconnected within the network, using the IMP and 50 K bits/s communication lines. The IMPs must operate completely independently of its local host computer and must continue functioning whether the host computer is operating properly or not. Furthermore, it must regard all the hosts in the network as having equal priority status. The IMPs act as message-switching centres, using store and forward techniques. They must transfer each message in under half a second, with an average of a fifth of a second for short messages. Although the message is transmitted on to other IMPs, it must retain a copy of the message until it is finally acknowledged. Each IMP consists of its own routing table, based on the communication delays the messages may be confronted with in reaching the required destination. These delays vary according to the queues and the

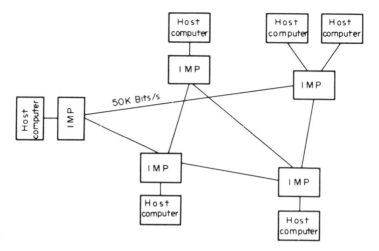

Fig. 10.6. The basic 'ARPA' network with its associated 'IMPS'.

connections available in the neighbouring IMPs. Each message is routed from IMP to IMP until it reaches an IMP that recognizes the destination address as belonging to its own host computer.

The ARPA network consists of another type of communication processor, which is known as TIP (abbreviated from Terminal-IMP). The TIP is also a communication interface, like the IMP, but it is intended to be a terminal flexible handling unit. The IMP is designed to handle the traffic from host computers into the network while allowing the terminals access to the network only through the host computers. This arrangement, however, requires all the remote terminals to be connected to the closest host computer, and since many terminals may use them only as a passing intermediate stage and will eventually overload the host computers, this would entail the need for extra hardware equipment to promote facility of access. The TIP can provide the access to the network for local or remote terminals where the latter are connected via modems. The TIP could also provide access to the network to a host computer in the same manner as with the IMP. A simplified illustration of the network showing both the IMP and the TIP is given in Fig. 10.7.

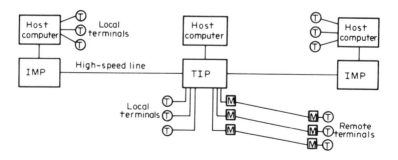

Fig. 10.7. Communication interface using both IMP and TIP.

The messages are transmitted between the IMPs in segments which are called 'packets'. Each packet includes a header with the destination address which is added by the IMP or TIP including the originator's number. When it reaches its final IMP, the packet is stripped of all the control information and passes the validated messages to the required host computer. The packets are routed dynamically from IMP to IMP, and their routes may be changed as they travel along the network. Efficiency is achieved by continuously updating the routing tables in each IMP.

The ARPA network is expanding throughout the U.S.A. and in addition includes access facilities in other countries, as seen in Fig. 10.8.

MERIT

This too is a distributed computer network similar to the ARPA, although the MERIT uses only standard voice grade lines with speeds up to 2000 bits/s. The results of this scheme may interest many companies in countries which have no access to high-speed communication lines, as it is designed around the general post office leased lines. Furthermore, it also provides dialed-up communication, which presents more system-operation flexibility and allows the enlarging of the configuration dynamically.

The MERIT network is abbreviated from Michigan Education Research Triad, which is a joint project of Michigan State University (MSU), the University of Michigan (UM) and Wayne State University (WSU). It is intended to create an educational computing network so that the universities can share their resources for both research and education. It is anticipated that this network will implement solutions to both hardware and software problems of network operation.

The theory of the system design is to encourage autonomy of operation of each of the computing centres while allowing their files to be shared when required by all the users in the other centres. The communication computers (CC) used in this system are the PDP 11/20, which act as an independent message-switching centre and provide identical interface to the communication lines, although each host computer retains its unique function.

The MERIT computer network is illustrated in Fig. 10.9. Each two communication computers are connected by a minimum of two full duplex communication lines. Extra lines may be provided by dialed connections, and for this purpose each communication computer consists of automatic calling units to set up these communication links when required. Each communication computer has been designed to provide easy transition to wide-band communication lines when the need arises.

DCS

This is a small experimental computer network based on the loop principle discussed in Chapter 8. The DCS is abbreviated from Distributed Computer System, and is intended to provide a low-cost reliable system. It is a flexible system where extra nodes may be added to the configuration. It utilizes in the configuration both mini and midi scale computers, and its potentialities could be of great interest. The network is under study at the University of California, at Irvine, under NSE funding.

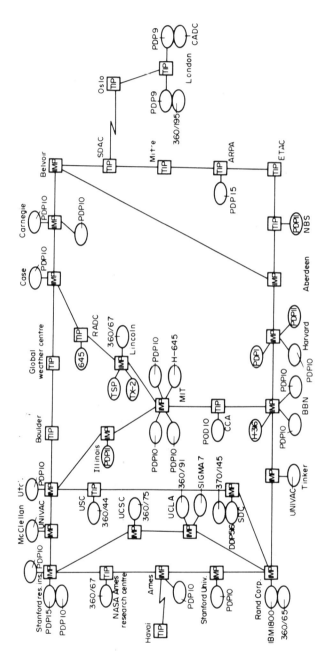

Fig. 10.8. Schematic logical map of the APPA network showing the approximate locations of the nodes and their interconnections.

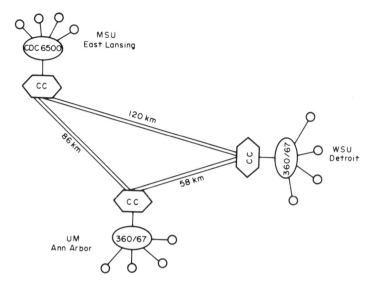

Fig. 10.9. The MERIT distributed computer network using voice grade lines.

The main difference between DCS and the ARPA networks is that it does not require a message-switching computer at each node but only a special hardware loop interface. The loop interface (LI) is what was described in Chapter 8 as station B. This feature is the main attraction of this system, as it requires little initial cost and is most flexible in permitting the additional stations. As such, it may interest many institutes which have no capital to finance large computing networks but still wish to share their resources.

The loop communication lines used in the DCS are Bell's TI type lines, which provide speeds up to 2.5 MB/s. A simplified illustration of this network is shown in Fig. 10.10.

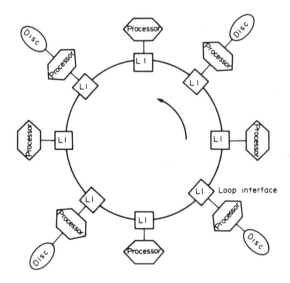

Fig. 10.10. The distributed computer system.

All the messages flow in the loop in a fixed direction and thus may reach all the stations without the requirement of special routing manipulations as in the ARPA network. This scheme also provides an easy system where messages may be directed to a number of stations. In the ARPA network the dispatching of messages to a number of destinations requires the duplicating of the messages and re-routing them by special action through many IMPs.

The loop operates on a time-sharing basis where time slots of equal size length are continuously flowing in the loop. The messages are placed in an empty slot and headed with the address of the destination processor. The loop interface, when it recognizes a message intended for its host processor, will lift the message off the loop and transmit it to its local destination. In this way all the messages are available to all the loop interface stations, which may accept or disregard them, depending on the destination address.

10.6 REFERENCES

1. Boan, T.H., "A standard for computer networks", *Computer*, Vol. 4, No. 3 (May/June 1971), pp. 10-14.
2. McPherson, J.C., "Data communication requirement of computer systems", *IEEE Spectrum* (Dec. 1967), pp. 42-45.
3. Amstutz, S.R., "Distributed intelligence in data communication networks", *Computer*, Vol. 4, No. 6 (Nov./Dec. 1971), pp. 27-32.
4. Henzel, R.A., "Some industrial applications of minicomputers", *Computer*, Vol. 4, No. 5 (Sept./Oct. 1971), pp. 7-12.
5. Ball, C.J., "Communication and the minicomputer", *Computer*, Vol. 4, No. 5 (Sept./Oct. 1971), pp. 13-21.
6. Townsend, M.J., "Communication control by computer – an introduction", *Telecommunications* (May 1972), pp. 33-38 and 60-62.
7. Fano, R.M., "On the social role of computer communications", *Proc. IEEE*, Vol. 60, No. 11, pp. 1249-1253.
8. Newport, C.B. and Ryzlak, J., "Communication processors", *Proc. IEEE*, Vol. 60, No. 11, pp. 1321-1332.
9. Decker, H., "Communication processing for large data networks", *Data Processing Magazine* (Nov. 1970).
10. Faber, D.J., "Networks, an introduction", *Datamation* (Apr. 1972).
11. Faber, D.J. and Larson, K., "The structure of a distributed communication system", presented at Symp. Computer Communications, April 4-6, 1972.
12. Roberts, G.L. and Wessler, B.D., "Computer network development to achieve resource sharing", *1970, Spring Joint Computer Conf. AFIPS*, pp. 543-549.
13. Heart, F.E., Kahn, R.E., Ornetsin, S.M., Crowther, W.R. and Walden, D.C., "The interface message processor for the ARPA computer network", *1970, Spring Joint Computer Conference, AFIPS*.
14. Kahn, R.E., "Resource-sharing computer communication networks", *Proc. IEEE*, Vol. 60, No. 11, pp. 1397-1407.
15. Matteson, R.G., "Computer processing for data communications", *Computer*, Vol. 6, No. 2 (Feb. 1973), pp. 15-19.
16. Tymann, B., "Computer to computer communication systems", *Computer*, Vol. 6, No. 2 (Feb. 1973), pp. 27-31.
17. Becher, W.D. and Aupperie, E.M., "The communication computer hardware of the MERIT computer network", *IEEE Trans. Comm.*, Vol. COM-20, No. 3 (June 1972), pp. 516-526.
18. Kahn, R.E. and Crowther, W.R., "Flow control in resource-sharing computer network", *IEEE Trans. Comm.*, Vol. COM-20, No. 3 (June 1972), pp. 539-546.
19. Davis, R.M., "Computer networks: a powerful national force", *Computer*, Vol. 16, No. 4 (Apr. 1973), pp. 14-18.
20. Pyke, T.N. and Blanc, R.P., "Computer networking technology – a state of the art review", *Computer*, Vol. 6, No. 8 (Aug. 1973), pp. 13-19.
21. Enslow, P.H., "Non-technical issues in network design – economics, legal, social and other considerations", *Computer*, Vol. 6, No. 9 (Aug. 1973), pp. 21-30.

22. Stafferud, E., Grobstein, D.L. and Uhlig, R.P., "Wholesale/retail specification in resource sharing networks", *Computer*, Vol. 6, No. 8 (Aug. 1973), pp. 31-37.
23. Kirstein, P.T., "Data communication by packet switching", *(IEE) Electronic and Power*, Vol. 19, No. 20 (15 Nov. 1973), pp. 503-508.

CHAPTER 11

Terminals and Displays

11.1 INTRODUCTION

Terminals and displays are the focal points of the Command and Control system and are provided to assist in better control of the organization resources. The displays are generally adopted in the upper levels of the organization, where they reflect a mirror picture of what occurs in the whole system. Terminals are used both in the lower and upper levels to provide means whereby data collected can be inserted into the system for feeding back to the display and to be available for retrieval on request.

As the terminals and displays serve as the essential links between the operator and the computerized system, their design is clearly of the utmost importance for the successful operation of the system. If it is well-designed structurally, a terminal or display can relieve the operator of many mental tasks, and if it is made aesthetically attractive, this feature will favour a readier acceptance of the whole system.

In Command and Control systems there are two basic areas where the interaction of human agent and machine takes place: (a) remote terminals; (b) the command post. Remote terminals are the points where the data is prepared in raw form for transmission to the computerized system and where the processed data is fed out. The remote terminals may thus be either input or output devices, or a combination of both. The command post is the point where all the processed data required for management decision is displayed, and accordingly it is basically an output device.

There are many devices which come under the title of terminals. They comprise those where input devices are: (a) keyboards; (b) readers; (c) switches; (d) function knobs; (e) light pens, and where the output devices are: (a) typewriters; (b) printers; (c) punches; (d) displays. There is no need to discuss these types of terminals, since they have been accepted in the computer industry as common input/output devices. Here we need deal only with those items which are important to the design of Command and Control systems.

In the last few years the terminal design has been extensively developed and its use has increased with the implementation of time-sharing systems. The major improvement has centred on the personal terminals which are used both in the command post and for remote entry terminals. These improvements have increased the man-machine interaction, where the most significant change has been the introduction of the CRT terminal instead of the conventional teletypewriter. Most terminals today in time-sharing systems use CRT displays with keyboard entry. A further significant advancement in this field is the introduction of a minicomputer within the terminal which completely transforms the traditional design approach to the terminal. The technological advances in this field have led to improved performance, increased capacity and a more aesthetic appearance of the terminal.

The command post is undoubtedly the centre of the whole system, being the place where the data is displayed and used for the supervisory control of the whole operation. Here, too, extensive development has been directed to the design of suitable large screen display systems, although the state-of-the-art has not yet caught up with the demand requirements. There is still much room for further work in this field, to improve the overall performance of the displays, and in particular in large screen graphical displays of real-time information status. This aspect will be discussed separately at the end of this chapter.

The design of the terminal is not only a technical but essentially a human engineering problem. A well-designed terminal, using all the latest techniques, answering to all the functional requirements and displaying the full information details required, may prove to be useless if the human reaction of the operator has not been taken into consideration to the fullest extent.

There is not only the question of selecting the most suitable terminal but also the matter of determining what and how much to display. Too many items displayed, although presenting the full required status picture, may be of little avail if the operator cannot grasp with a single glance the important items which he must deal with. In some terminals, in order to improve performance, words are inserted seriatim at a speed suitable for human reaction. This solution is unsatisfactory, however, as there the operator must keep his eyes glued to the display terminal for fear he misses an important item, and so becomes soon tired and unable to operate such terminals for prolonged periods. Other terminals use flashing lights to indicate a new item requiring the operator's immediate action. What happens, however, when there are a lot of new items which must be displayed before the operator can get the chance to respond? Should the whole display flash? There is no doubt that these points could be overcome by carefully designing the amount of data to be displayed and the rate of data change. At any given time only a minimum amount of data items should be displayed, and no fresh data should be inserted on the display panel before the 'old' data has been signalled as accepted by the operator's actions. The data displayed should include as little routine information as possible, concentrating rather on the ambiguous data which require the operator's immediate decision and intervention.

The introduction of colour into the displays is a most convenient form for increasing the data volume where a multitude of different tasks can be presented together. Apart from this, soothing colours are more acceptable than flashing lights and are less of a strain on the eyes, although they might not be equally effective. The colour schemes and the flashing lights must be designed by a specialist in human engineering, who may need to modify the design requirements defined by the system engineer.

The terminals, whether situated at the remote stations or at the command post, must be simple and easy to manipulate. There should be no need for lengthy training courses to learn how to operate these terminals; indeed, the design should be aimed at the unskilled operator. What is required above all is that the terminal functions should be logical, so that all the workers associated with the operation will be able to handle the terminals. An additional feature is having the functional keys well labelled and self-explanatory, thus obviating the need for special training. The terminal must always respond clearly and positively to all the operator's commands, that is, by visually displaying an acknowledgement to the commands. If any of these commands are forbidden or unclear and if looking at the keyboard does not give sufficient information, then special lights should indicate what is amiss.

Although most of the terminals used in Command and Control systems are intended for the unspecialized operator, it is to be assumed that he has a good enough background knowledge to be able to carry out the special functions after a little training. A more basic programming training will be needed, however, for some terminal applications, but even here the terminals should be simple and easy to operate. The language used to communicate with the system by means of the terminal must be as close as possible to the natural language, with limited number of abbreviations. In many applications the computer can take the initiative in the dialogue, such as where it asks the operator direct questions and the operator only has to give short simple responses. Other configurations provide the operator with ready-built tables, where the operator is only required to enter the data in the specially provided blank spaces.

Many of the operator's functions are routine and likewise the communication functions. Where the functions can be performed by the terminal, both the operator and the centralized computer could be relieved of repetitive tasks. Such terminals are referred to as 'smart' or 'intelligent' terminals, and will be discussed in Section 11.3. With them the data required for transmission can be displayed, thus enabling the operator to check and correct his message before it is dispatched. This subsequently requires temporary storage in the terminal itself, including a small amount of processing. In fact the 'smart' terminals contain a minicomputer which proves to be a boon, since it saves central computer time, operator's time, communication time, and prevents errors being introduced into the system.

11.2 REMOTE TERMINALS

Remote terminals connect the system users scattered over a wide geographical area with the centralized system by means of communication channels. The term 'terminal' includes a wide variety of computer input-output peripheral equipment which may be in the form of teletypewriters, CRT's, keyboards, punch readers, etc. For Command and Control applications, the terminals may be divided into five major categories according to their fields of application:

(a) Data-acquisition terminals.
(b) Control terminals.
(c) Data-transaction terminals.
(d) Enquiry terminals.
(e) Display terminals.

In the class of data acquisition terminals are the sensor type terminals which are "hard wired" into the process so as to collect on line real-time information. These terminals generally contain measurement interface equipment, A/D converters, storage buffers, reference clocks, communication multiplexors and modems. In this category one can also include all the types of automatic data collection, such as badge readers and remote telemetry sensors. Due to the wide variety of the applications that may require data acquisition terminals, these terminals are generally custom designed for each application.

The control terminals are complementary to the data acquisition terminals, since they are intended to automatically implement the system decisions. These terminals contain instrument interface equipment, D/A convertors, storage buffers, communication demultiplexors and modems. They are used in the feedback process of the system where the operation must change its control path in real time. An example of such an application can be found in the Command and Control system required for electricity and water supplies.

Corresponding to the data acquisition and control terminals are the data-transaction terminals. The data-acquisition terminals discussed above are of the instrument type where the data is read automatically, at given intervals or whenever there is a change in any of the input conditions. In data-transaction terminals, the data is collected and inserted by human operation instead of by automatic instruments. As with the other terminals, the data-transaction terminals are connected on-line to the system, that is, the data which is inserted by the human agent is transferred directly to its destination, receiving instantaneous reaction and providing the operator with a reply in "real-time". Nevertheless, the instantaneous real-time reaction required for transaction terminals is much slower than that required for control terminals. The real-time element here depends on the human acceptance of presented results. In the category of transaction terminals are all specially programmed instruments operated by humans, such as those used by the agents for air seat reservations. With these terminals, the operator is confronted with a standardized format terminal where he inserts his information by selecting specific functional keys. The proper design of the transaction terminals provides an easy system interface, where the unskilled operator can transmit complicated information.

Data enquiry terminals are the most widely used terminals, particularly for time-sharing systems. In contrast to the previous type terminal discussed, these terminals need not be custom designed for a specific application but may be mass produced to cover a wide range of applications. With these enquiry terminals, the operator can insert his specific request and receive the computer's reply on the same terminal. The operator is free to use any format which may be convenient to his application, for the format is controlled only by software and not by hardware (as in transaction terminals). However simple and easy the handling is made, the operator must be trained to insert the data into the correct format so as to enable the computer to follow instructions and perform the necessary actions. As already explained, the system may be designed to assist the operator using these terminals by asking specific questions or by instructing him to insert the required information.

All the enquiry terminals must act as both input and output devices so as to provide the operator with a single terminal where he can transmit and receive his messages. Furthermore, these terminals can provide the operator with a copy of his questions and the computer's answers, with the copy in one of two forms, either hard in the form of printed data, or soft in the form of a visual picture. The earlier enquiry terminals provided only hard copies which proved to be slow and tedious processes, whereas today the tendency is to use terminals which provide only a soft copy and are simpler to handle and more convenient for prolonged conversation with the system. The simplest enquiry terminals are the teletypewriters, which provide hard copy and are thus generally suited for long programme transactions with a lot of data entries. More sophisticated enquiry terminals are the CRT/keyboards, which provide only soft copies and are more suited for the unskilled operators. With these terminals, short messages and their replies can be displayed in front of the operator in an easy readable form.

The soft copy visual enquiry terminals increase operator efficiency. In fact much development has been directed in the last few years to improve the performance of the enquiry terminals, but there is still room for further improvements. These developments have been directed towards the improvement of both the visual and the logical performance of the terminals. The most significant advances are in adding colours, using a light pen and incorporating a minicomputer within the terminals. Much development has been centred on the external appearance of the terminal so as to make it aesthetically suitable to the management board room. Colour displays add new characteristic qualities to the performance of the terminals by enabling different items to be indicated by different colours. The light pen provides very easy and speedy access to all the items displayed. The inclusion of the minicomputer within the terminal increases its logical power, as it can operate as a stand-alone unit in solving some simple programme problems.

The use of visual displays for enquiry terminals is increasing in popularity over teletypewriters and other similar type of terminal. Nevertheless, there are many applications which may require not only a soft but also a hard copy. The technological development of a combined soft and hard copy in a single unit is by no means complete, and there is room for development of a CRT terminal which can print out the display contents when required. This "print-out" equipment must be produced within the terminal, so as not to take up too much room and be simple to operate. Furthermore, it must be able to operate noiselessly and at the same speed as the CRT.

Display terminals are primary output devices where the data is displayed for decision making in the command post. They may be divided into two main groups: personal displays and large screen displays. Personal displays are of the type of CRT/keyboard displays already discussed. Among them one must also include the graphic displays which are used for computer-aided designs, which allow on-line graphical conversion for sophisticated analysis of problems. In general, when referring to display terminals one really indicates the second group, which consists of large screen displays. These terminals use completely different technologies from any of the other types of computer peripherals, and furthermore they require a separate computer to operate them. These techniques call for individual discussion. It should be added that the state of the art of these displays is still at a stage where the systems available do not cover all the application demands.

11.3 DISTRIBUTED INTELLIGENCE

The terminals are used to collect the data and transfer it to the central processor, which in theory should concentrate on processing the data. In practice, however, apart from the task of processing and editing messages, most of the work of the central processor, in a distributed network, is concerned with routine tasks required for implementing the communication between the centre and the remote terminals. In addition, the central processor has to regulate the data to be displayed or printed by controlling the traffic between the output terminal and centre. One of the ways for reducing the routine load from the central processor is to transfer most of the communication tasks to the front-end processor, as discussed in Chapter 10. This, however, still leaves the bulk of the routine communication tasks in the centre and does not reduce the traffic volume along the lines themselves.

The system design policy today is to relieve the centre of all the tasks not directly associated with data processing and transferring as much as possible to the terminal. This technique, which endows the terminals with the power of performing intelligent functions, is referred to as "distributed intelligence". Justification for distributed intelligence is to be found in the improvement of the system operation efficiency and the reduction of the traffic volume flowing between the terminal and the centre.

The distributed intelligence is achieved by introducing a minicomputer within or near the terminal. A built-in minicomputer within the terminal presents the local operator with maximum facilities without the need to refer to and disturb the central processor. Such terminals are characterized by the feature of being able to function as a stand-alone unit, for which reason they are referred to as 'smart' or 'intelligent' terminals.

The terminals in time-sharing systems are generally of the enquiry type, i.e. they are used as data entry where the messages are constructed by manual operators. Much of the conversation between the terminal and the computer involves editing the message. The intelligent terminal provides the operator with an instrument with which he can visually see the message as he constructs it on his personal CRT, correct the details by deleting any erroneous data or add any further information. In the absence of local logic, these facilities would require the terminal to be connected to the computer throughout the full length of time the message is being constructed. The use of intelligent terminals allows the message to be constructed "off-line" and to be transmitted only when ready. Furthermore, these terminals can then transmit their data at the maximum possible speed (up to 9600 bits/s) and not be restricted to the human speed of constructing the message as with the 'old' teletypewriters. Some intelligent terminals may incorporate many more sophisticated processing abilities without disturbing the central computer. An example of such processing facilities is in the preparation of fixed tables with the appropriate headings. The tables could be designed with restricted field entries, allowing some with only numerical data while others with only alpha characters. Some tables would consist of a field restricted to the operator, and others for the computer's reply. In fact, there is no limit to the facilities that can be presented by such terminals, provided the memory capacity available in the terminal can grant it.

When a minicomputer is built into the terminal itself, it offers the maximum facilities to the operator, but its adoption is not justified for all Command and Control applications. For job-entry terminals it is advisable to use a separate minicomputer with each terminal and to include a CRT display (as shown in Fig. 11.1). With this type of

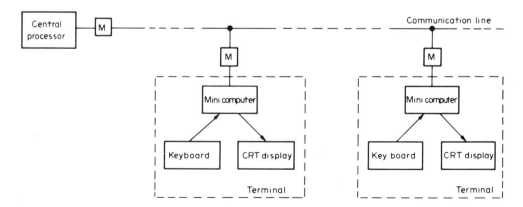

Fig. 11.1. Stand-alone intelligent terminals.

terminal, the operator can prepare his message on his CRT display and then transmit it as a whole message or a line at a time, after he has made sure that the information is complete and error free. The terminal can also be provided with a printer if the operator requires a hard copy for later reference of his transmitted transactions. The flexibility of such a terminal is great and its functions depend only on the programming design with which it is fed, as is the case of any other minicomputer. On the other hand, the storage capacity of these terminals is small (generally only 4 K) and consequently the number of facilities is limited by the memory space available. Today, many manufacturers provide a CRT terminal with a built-in minicomputer as described above. The central computer, in a system using these terminals, can address each unit separately, while the terminal is provided with enough software facilities to recognize such a call.

For standard type inquiry entry terminals, such as those required for airline reservation, it would be wasteful to provide each individual terminal with local intelligence. Even if the data is to be displayed on a CRT, the amount of editing is limited and so there is no justification for a separate minicomputer with each terminal. However, even in this type of application there is room for distributed intelligence whereby the load of the communication facilities is removed from the central processor. In such systems the distributed intelligence is transferred to a local concentrator which controls a number of terminals (as shown in Fig. 11.2). The concentrator is associated directly with each terminal, using separate lines, while it has only one line to the central processor. The central processor can still address each terminal separately, although the actual selection transaction of the terminal due to a poll request is performed by the concentrator. The concentrator acts as a multiplexor, collecting the messages from all the terminals and sending them to the central processor. It also provides each terminal with some minimum logic facilities. Furthermore, it reduces the traffic to the central site (though not between the terminal and concentrator) by editing the messages into a standard format which may have been generated from a variety of terminal configurations. The central processor can then regard all the terminals as looking and behaving alike, without the need for special programmes to handle each terminal type separately. The task of recognizing the terminal type is thus transferred to the concentrator, thereby relieving the central processor of this task.

As displays offer many advantages over the TTYs, they tend to replace them for most time-sharing applications. However, as already mentioned, not all the applications require a built-in computer, even though a concentrator is not sufficient to provide all the

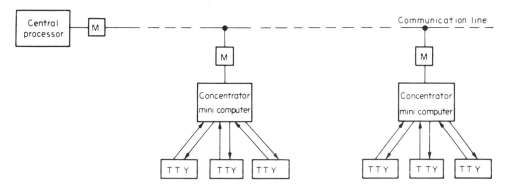

Fig. 11.2. Simple distributed intelligence.

necessary 'intelligence'. A number of 'dumb' CRTs can be clustered together (as shown in Fig. 11.3) with a computer controller providing each terminal with the required intelligence.

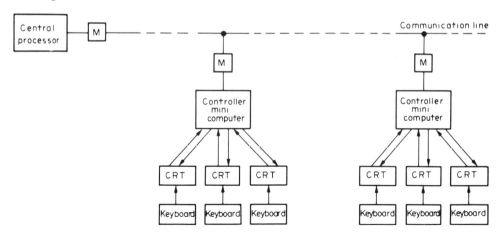

Fig. 11.3. Clustered CRT terminals in a simple distributed intelligence configuration.

Although the initial cost of distributed intelligence may seem relatively high, the considerable saving in the system efficiency justifies its adoption. Most of the gain comes from the central site, which is freed to perform many more processing tasks. Further economy is offered in the communication network, where fewer channels are required as a consequence of the reduction in the traffic routine. Full economy is achieved where the maximum possible operations are performed locally.

With intelligent terminals, each individual unit is uniquely identified throughout the system, with no need for direct communication lines to the centre. Furthermore, the system can always have a complete real-time knowledge of the terminal's status, since it can be made to report its condition automatically. The terminals can identify themselves and transmit any status information, with no need for human intervention. Where no free links are available to the centre, the terminals can retain the data and wait for the line to become free, without the need for any further human instructions.

There are terminals which consist of a wired-in read-only memory which permits only minimum communication control functions according to a predetermined programme. Although these terminals are not considered to be 'smart' terminals, they do help in distributing the intelligence.

11.4 LARGE-SCREEN DISPLAYS

In the command post of a Command and Control system some form of display is needed to provide a fair-sized audience with status information required for decision making. The large screens adopted for this purpose are designed to ensure rapid access to dynamically changing data presented in real-time. The highly sophisticated nature of the data makes it essential for it to be displayed in colour, with each item clearly distinguishable and all factually accurate.

The screen has to be large enough to exhibit the full system's status, that is, both the static and the dynamic information required for the construction of the situation status. The static information generally consists of items which seldom change, such as maps, plans and tables. The dynamic information demonstrates all the items which require immediate attention and comprises all the updated changeable data entries. In other words, the displayed dynamic information consists of all the on-line measurements exhibiting the sudden changes and the deterioration of the readings. Although the decision making is based on the dynamic information, it cannot be used without reference to the static information. A simple example of such an application can be seen in flight control, where the display of aircraft movement is not informative unless it is superimposed on a map.

Large-screen displays are available in many forms and range from electro-chemical character displays to colour television projection displays. It is not possible here to describe all the available systems, but it may suffice to give a general indication of the latest development.

Large-screen displays can be divided into four categories according to the applications they are intended for, viz.

(a) Those required for the display of tabulated alphanumeric data, as used in airports for the display of arrivals and departures of flights.

(b) Those required for the display of graphic data, as used for computer-aided design and for the demonstration of production curves in the management room.

(c) Those required for the display of moving targets along fixed paths, as used in train control or electricity supply.

(d) Those required for the display of moving targets along a random course, as used in air-traffic control.

All these four types may be used in Command and Control systems and may be required for the decision making in the command post. The most widely used display is that given by the first category, although the dynamic information is better displayed by the other three categories. The techniques for the first three categories are well established and will not be discussed here. The fourth category is of particular interest since it is intended for the very large sophisticated Command and Control systems. However, with the state of the art today regarding this type of display, not all the design problems have been solved, and there is still much room for further development.

The requirement to display on-line random movement of targets on a map background is becoming an important item in Command and Control systems. Many examples can be given, such as flight control, satellite control and mobile traffic control. In mobile-traffic control applications, the tendency had once been to concentrate control of the traffic along a particular route, but it was soon found that the traffic density on one route affects the traffic along other routes. This point is particularly relevant to traffic flow in a large city. Here the traffic jams are not only related to each other but their location may vary according to the time in the day. Such a Command and Control system is controlled by a computer which operates all the traffic lights, but, nevertheless, the command post must have the facilities for supervising its operation. In other words, the command post controlling the city traffic must have direct access to the data relating to all the road congestions, and display it on a map. It is obvious that a single map will not suffice and that an assortment of maps are needed. In effect, the large screen in the command post must be able to display a number of maps of various sizes and scales, selected according to the immediate need. It should be appreciated that these maps only provide the static background data, while all the dynamic data, showing the moving and changing items, must be superimposed on the map in real-time. The operator must be capable of selecting any area on the map and enlarging it so as to pinpoint a particular item. When doing so, all the relevant dynamical data should be expanded according to the associated updated computer file. Furthermore, the maps (with all the relevant real-time data) should permit their being rolled up or down or sideways without leaving gaps between adjacent map areas.

The requirements of the sophisticated large screen display, as specified for the above example, makes its implementation possible only if the maps are kept in the computer memory in digital form. That is, the computer memory holds both the static and dynamic information and presents them as one dynamic display. It can be inferred that the number of details for the maps, including the colours, is considerable, and this makes the required storage capacity and the time necessary for displaying them, quite a problem. Furthermore, displaying such a complex picture on a TV screen requires far more than the available resolution of 1000 lines. In fact, the writer does not know of any system which on request can display electronically a large number of geographical maps together with all the dynamic details. The solution generally used today is to display each map as a separate item and then superimpose on it the required dynamic details. The main problem is to ensure that all the details overlap each other in a manner which is termed the registration of the picture. The current state-of-the-art can present a registration in the range of 0.1–1%, which is not good enough for sophisticated dynamic map displays. One accepted solution is to use a synthesis map presenting only the bare necessary items.

The general technique used today for displaying pictures is achieved by separately projecting the dynamic and static items on to a screen. The dynamic data is fed from the computer and displayed with the aid of a television raster while the static data is displayed by a slide projector. This by no means presents the best or only available solution, and much research work is still being carried out in this direction.

The items presented on the display screen must have the same resolution when viewed at a distance or close by. The operators may want either to be in close contact with the screen so as to be able to measure distances of items with a ruler, or to remain at a distance in order to get the feeling of the full picture. The operators must be careful to cast no shadows on the screen or prevent other people from viewing when he is near the screen. This can be achieved only if rear projection is used. An important point to consider is that the rear projection does not affect the operator's health through any stray rays.

A large number of facilities are required to support the large screen display, as can be seen by the general layout illustrated in Fig. 11.4. The display system is completely separate from the rest of the Command and Control organization, to allow it to operate as a stand-alone project. Nevertheless, it must be connected on-line to the central processor of the system so that it can receive all the processed up-dated information. The computer interface connecting the two systems may require to be connected via modems since in most cases they are not situated in the same location. A special processor is required at the display site, to control all the internal operations without the need to refer back to the central processor. These local control operations are programmed separately, to allow for the editing of the information displayed by means of the facilities, such as blinking specific items, adding and deleting data, etc. The display system for these control applications is provided with an operator's keyboard whereby the operator can communicate with the system.

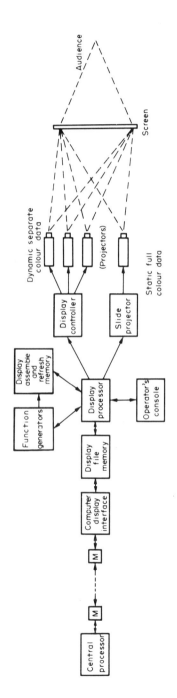

Fig. 11.4. General plan of a large-screen display.

The display system consists of three memory units. The first is the main memory which stores all the data files selected for display and keeps them ready for quick transfer from one status picture to another. The second is the refresh memory, where the data is assembled before it is displayed and the complete display is refreshed. This stage is necessary because most CRT type displays do not retain the image, and to avoid any flicker of the items displayed the picture must be refreshed at a high repetition rate. The third memory is of a read-only type, which is known as the function generator. This unit generates all the standard symbols by converting the computer binary information into displayable alphanumeric and other accepted symbols.

In the system shown in Fig. 11.4 the static data is displayed in full colour by means of a slide projector which is controlled by the display computer, while the dynamic data is superimposed by separate projectors, one for each colour, to indicate the data held in the refresh memory.

An alternative method for displaying the static data is to merge it with the television image with the aid of a video switch matrix (as shown in Fig. 11.5). The maps or

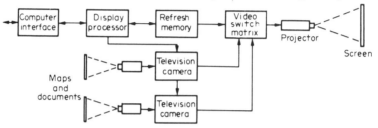

Fig. 11.5. Multiplexing static and dynamic data.

documents are deflected by a television camera with the video output of the camera multiplexed with that received from the computer. The advantage of such a system is that the registration is performed electronically. In addition, it can superimpose a number of static pictures simultaneously with that of the dynamic picture.

11.5 LARGE-SCREEN DISPLAY TECHNIQUES

From the previous section it can be gathered that there is no single design technique for large displays, and that the techniques available do not meet all the application requirements. Many large display techniques have been discarded after prototype evaluation, while many others are still in the R & D stages. A few of the systems available on the market are presented here for consideration.

The simplest electronic display is one built of a row of segmented gas-discharge units, each displaying a character by selecting the appropriate segments. This type of display, although widely used, is not suitable for the command post, because it is generally too small for displaying the full information content, let alone the background static picture. The tendency today is to employ a dot matrix rather than a segment array. This allows more types of symbols to be displayed by the same array and in a more accurate form. (An example of such an array, using gas-discharge dot matrix, is given in Fig. 11.6). The initial matrix size required for the display of each symbol or character is 5 X 7 dots, but this is being increased to a matrix size of 9 X 11 to exhibit better human engineering properties. It should be noted that the smaller matrix (still widely used) leaves room for many ambiguities, such as confusing B and 8.

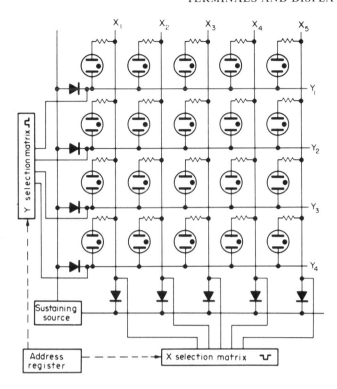

Fig. 11.6. Gas-discharge dot matrix.

A more advanced form is the plasma panel, based on a dot matrix of gas-discharge cells. The characters are inserted on to the panel by selecting suitable dots in the matrix. This is performed by the simultaneous driving of the horizontal and vertical lines of the matrix (i.e. anode and cathode lines) in a similar mode to that used in cross-bar systems. The anode lines are selected with a high voltage (250 V) simultaneously with a priming ionization at the cathode. A thin layer of dielectric isolates the vertical and horizontal lines, which retain the charge in the cross-point after selection. An alternating voltage, of a sufficient potential, ionizes the charge cells so as to emit light. Panels with over 15,000 cells are available for graphic applications, although much larger panels, with over a million cross-points, are under development. In some applications, the plasma panels displaying the dynamic data are mounted in front of a translucent screen upon which static data from slides may be projected.

A similar display uses a matrix of light-emitting diodes (LED) where a pair of cells closely mounted exhibit two colours. This gives the impression that the colours originate from the same point. Since the diodes have no internal memory, the retention of the data is performed by separate memories, an arrangement which increases the cost. Such panels, which have the advantage of being flat and operating with low voltage, are used only for small alphanumeric displays.

Another similar technique under development is that of the liquid crystal, in which a potential applied in the matrix lines causes the alignment of the crystals in the cross-points. By application of an external light source, the crystals in the liquid may either transmit the light through the panel or scatter it.

The three display techniques as described are limited in size, as each dot requires a specific cell. Further advancement has been made by using television cathode-ray tubes (CRT) which can provide more facilities, with an increasing amount of digital data and with multi-colour display. The static data may be superimposed on the main beam of the dynamic data by rear port projection on to the tube face (as shown in Fig. 11.7). Nevertheless, the current state-of-the-art of these tubes has many unsolved problems, of which the most vital one is that of the amount of details displayed with the required resolution and registration.

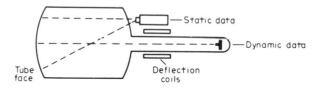

Fig. 11.7. CRT display with rear port projection.

Television tubes are limited in face size, and this limits their application for large-screen displays. Nonetheless, most of the advanced large-screen displays which are based on projection techniques are in fact an extension of the CRT techniques. In these applications, the picture image on the tube is magnified and projected on to a screen by the combination of a spherical mirror and an optical system. In one of the systems (shown in Fig. 11.8) the modulated television image generated by the cathode is deflected

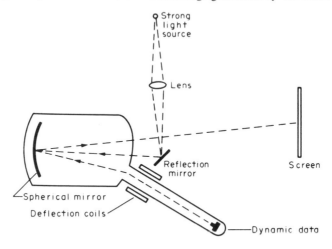

Fig. 11.8. Large-screen display based on CRT projection techniques.

by a spherical mirror situated at the back of the tube face. This mirror is composed of a thin coating of viscous liquid which is deformed by the electric beam charges. A light generated from a strong source is directed to fall on to the same spherical mirror with the aid of an optical lens and another mirror. This light beam is defracted, according to the image deformation, and is projected back in to a large screen.

Similar techniques are being developed where the CRT tube face itself acts as a spherical mirror, whereby it reflects forward the displayed image on to a large-screen situated in front of the tube face, as shown in Fig. 11.9. As in the previous technique, a

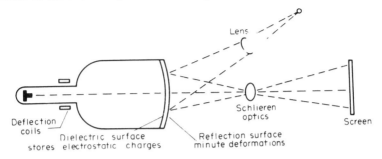

Fig. 11.9. CRT forward projection technique.

strong light source is deflected on to the tube face, which consequently is deformed according to the picture inserted on it from the other side. The tube surface is coated with special non-transparent oxides which are susceptible to electrical charges, causing crystal deformation which consequently scatters the light which is projected on it from the other side at a particular angle.

The brightness achieved in both projection techniques is a function of the power provided by the light source. This is also the main drawback of the system, for they require special ventilation conditions, as the light source converts many kilowatts of electricity power into heat. It should be noted, however, that forward reflection requires less power than backward reflection.

As already mentioned, large-screen displays based on CRT techniques are limited by the television resolution of 1000 lines. This limits the amount of detail possible and the registration between the items. There are other earlier techniques which do not suffer from these limitations (although they are subject to limitation in displaying real-time dynamic data) but they are still nevertheless unsuitable for many Command and Control applications. Examples of such techniques are the film projection and the slide scribing projection. In the scribing projection technique the data is inserted on to a slide in the manner in which an X-Y plotter operates. The moving targets are inscribed on to the surface of the slide, which is coated with a special non-transparent substance. Thus, wherever data is inscribed, light may penetrate and be projected onto the screen. The basic surface of the slide is transparent, although slides of different colours may be employed. The data may be inserted in real-time but once it is on the surface it is difficult to erase. Although the dynamic data inserted on the slides may be removed, there are no means of erasing particular items except by the relatively slow process of clearing the full dynamic image from the slide. Dynamic data can only be modified by changing the slides, though this would require the re-inscribing of all the dynamic details still currently valid.

11.6 REFERENCES

1. Hobbs, L.C., "Terminals", *Proc. IEEE*, Vol. 60, No. 11 (Nov. 1972), pp. 1273-1284.
2. Salzman, R.M., "An outlook for the terminal industry in the United States", *Computer*, Vol. 4, No. 6 (Nov./Dec. 1931), pp. 18-25.

3. Amstutz, S.R., "Distributed intelligence to data communication networks", *Computer*, Vol. 4, No. 6 (Nov./Dec. 1971), pp. 27-32.
4. Kamman, A.B., "How to pick CRT terminals", *Data Processing Magazine* (Apr. 1971).
5. Moll, A.P., "Data display for real time telemetry", *IEEE Trans. Comm. Tech.*, Vol. COM-14, No. 6 (Dec. 1966), pp. 843-848.
6. Rosenthal, C.W., "Increasing capabilities to interactive computer graphic terminals", *Computer*, Vol. 5, No. 6 (Nov./Dec. 1972), pp. 48-53.
7. Standeven, J., Mihailovic, S. and Edwards, D.B.G., "Multiconsole-computer display system utilising television techniques", *Proc. IEEE*, Vol. 115, No. 10 (Oct. 1968), pp. 1375-1379.
8. Grochow, J.M., "Real-time graphic display of time-sharing system operating characteristics", *AFIPS*, Vol. 35, Fall Joint Computer Conference, 1969, pp. 379-386.
9. Machover, C., "Display terminal-status and standards", *Data & Communication*, Vol. 2, No. 4 (Sept./Oct. 1974), pp. 8-11.
10. Bryden, J.E., "Visual display systems", *Telecommunication* (May 1972), pp. 22-31, and in *Electronic Progress*, Raytheon Co., Vol. XIII, No. 3 (Fall 1971), pp. 2-10.

Error Control

12.1 INTRODUCTION

A fundamental requirement of any Command and Control system is that the message received at any node in the system should be without errors. Unfortunately, the communication network is not impervious to noise, and that is liable to cause distortion of the transmitted information. To prevent errors entering into the text of the messages, means must be inserted into the system, first to detect the presence of the errors at the particular position and then to find ways of correcting them. As the system and its associated communication network become more sophisticated, the requirement for error-free transmission becomes increasingly important.

In any data-transmission medium, the data received must convey some measure of assurance of its accuracy. For this purpose, redundancy is usually inserted into the data stream transmitted over the communication links. This has the effect of reducing the relative volume of the effective information bits originally transmitted. But it also causes delays, and increases the system cost arising from the complexity of the equipment needed to detect and correct the errors. The design aim of any system is therefore directed towards maximum transmission rates without the cost and the error rates becoming too excessive.

Means of error control are introduced into all data-transmission media to solve the two major problems of error detection and correction. The source of the errors is generally the noise in the line, especially the impulse types. These errors are created randomly with no fixed pattern, although most of them tend to occur in clusters. It must be noted that the transmission may at some times be relatively free of interferences, while at others it may be considerably distorted. It is the unpredictable behaviour of the interference that makes error control a difficult problem. Indeed, error control cannot effectively handle the errors from a random statistical formula, as the fact must be accepted that the errors appearing in clustered form are due to a single noise burst.

Various techniques for the detection and correction of errors have been discussed in the technical literature; but it is beyond the scope of this book to deal fully with them. What is offered here must be considered as the introduction of the subject, with a description of the major trends in the field. When designing any data-transmission system, error control of some kind must be allowed for. While error correction may not always be essential, its detection is imperative. In some system applications a number of error-control techniques are used simultaneously, one to detect random errors and the other for bursts of errors. It is also common in many systems to use double checks, one on each character and another on each block. It is questionable whether the use of a number of techniques increases the system's reliability, and the designer should consider whether the extra cost will truly improve the system's efficiency. No error-control system can be 100% perfect, but the final error rate can certainly be improved by several factors, depending on the amount of sophisticated equipment and the redundancy the designer is prepared to incorporate into the system. It is worth adding that the percentage of redundancy is by no means related to the transmission reliability.

The techniques to be used for a particular system depend on the application involved and the performance required. Furthermore, the characteristics of the links play an important part in selecting the techniques. Dialled lines are generally noisy, while better results can be achieved with private lines. Leased lines, if not so specified, are considered noisy, as they still have to pass through the exchange. Radio links are generally less susceptible to noise, although they do suffer from fading. In each application the designers must define the accuracy they wish to accomplish, determined by the level of errors which may be allowed in the system and how much the designer is ready to pay to reach this error-free level.

Before selecting a particular error-control system, the type of the data transmitted must be defined. Where the information bits are continuous, only convolutional error codes can be used, while if the information may be grouped together, block error codes may be adopted. For most Command and Control applications, block codes are suggested, but not for telemetry applications, which require convolutional codes. Block codes generally only require error detection (although there are many block error correcting codes) while convolutional codes commonly require error correction.

When selecting the suitable error-control technique, the designer should consider the following points:

(a) The type of the data to be transmitted.
(b) The degree of accuracy required in the received data.
(c) The inherent reliability required.
(d) The number (or per cent) of incorrect digits or messages which may be allowed to be let through.
(e) The delays allowed in the system.
(f) The accepted redundancy of the data allowed (the volume of the data transmitted versus the accuracy required).
(g) The required throughput in the system (the throughput is reduced by redundancy, coding delays and requests for retransmission).
(h) The type of links available and the interferences which may be anticipated in the links.
(i) The efficiency of the data transmission on the communication links.
(j) The implementation cost and the cost efficiency of the various error-control techniques.

The technique selected should be simple enough for it to be implemented with the minimum of equipment and capable of achieving low incidence of delays and ensuring that the number of check digits used relative to the number of information digits is small.

Errors are detected by the additional insertion of redundant information into the data transmitted. This, however, is not always sufficient to indicate multiple errors. In more sophisticated systems there is the further requirement of a quality detector, which has a better chance of indicating severe interference which could remain undetected by simple parity checking. A general diagram of an error-control system is shown in Fig. 12.1.

The detection of errors is usually at the receiving end, while the correction of the

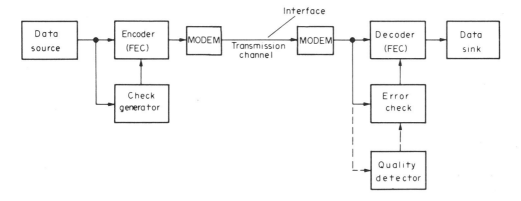

Fig. 12.1. General diagram of a forward error-control system (FEC).

errors may be done either at the receiving end or by a request for retransmission. Despite the vast amount of theoretical studies that has been directed towards error-correction codes, their applications are still rather limited. The most commonly used technique for the purpose is that of retransmission. Wherever circumstances permit, correction by retransmission is preferable. This technique requires that messages be held at the transmission point or at any intermediate point in case a request for retransmission is received. This is to ensure that no information message is lost in the system. In some applications retransmission may not be in use and, there, either the data is automatically corrected or only the fact that there is an error is recorded. The latter alternative may be the case in telemetry, where statistical data is gathered.

Requests for retransmission may be made either on a separate channel or on the same channel, and occur during the intervals between the blocks. In systems where no retransmission is used, there is no need for a return path and then they are referred to as forward-error control (FEC). The retransmission error-control technique, shown in Fig. 12.2, is commonly referred to as automatic-repeat request (ARQ).

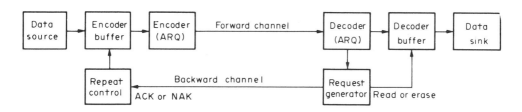

Fig. 12.2. Automatic repeat request error-control system (ARQ).

The communication procedure with retransmission between any two nodes in a network is as follows. (A node in this case may be a terminal, concentrator or computer.) The sending node transmits the information block but still retains the text while awaiting an acknowledgement. At the receiving node, the block is checked for errors by comparing the check digits with the information digits. The receiving nodes can then accept the block by means of an acknowledgement code, request a retransmission by sending a no-acknowledgement code sign or by simply ignoring the block. Acknowledgement is performed by sending along in the return path the accepted character label of 'ACK'; no-acknowledgement is recognized by sending the character label of 'NAK'; but of course no signal at all is returned when the received block is ignored. As the cause of a block being ignored may be due to the line being down or to the receiving node being unable to accept the block, the sending node waits a fixed period after sending the block, and only if no reply is received from the second node will it then look for alternative routes.

Retransmission is generally limited to three times, to prevent both sending and receiving nodes being continuously engaged where the line is faulty. A sequence of three NAKs or the absence of a reply from the receiving node will cause an alarm in the system at the sending node. This alarm must cause the investigation of the particular link before it may be used again.

Repeated retransmission of the same block three times on a noisy line may present three different blocks. As the chances are small that the errors have occurred each time in the same bit positions, in most applications these three blocks may be held and compared individually. The received message could then be reconstructed by majority decision. Although this technique is by no means perfect, it may be the best solution for applications where there are no alternative routes available.

In a distributed computer network, the block is transferred from node to node until it reaches its final destination, as shown in Fig. 12.3. The block is held in each node till a

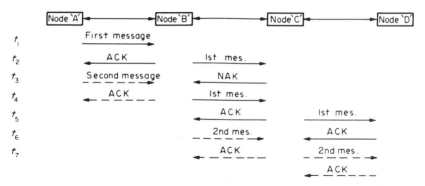

Fig. 12.3. The transfer of messages in an 'ARQ' control system.

positive acknowledgement is received. A second block cannot be transmitted on the same link till a positive acknowledgement is received for the previous block. At each node, the hardware inserts new terminating framing characters and error-checking characters to each transmitted block. These characters are recognized by the next node and are used for checking the transmission only on that particular link. When the message is retransmitted to the next node in sequence, a completely new set of characters must be added. Each node in turn stores the block till it has received an acknowledgement from the following node in sequence, but it does not hold on to it until it reaches its final destination.

The earlier error-control codes were based on parity check codes which were built within the character. However, since the noise is generally clustered in a burst of impulses, the likelihood of distortion of both the information digits and the parity digits is great. A much better error-control technique is to have the code within the block, where one character or a number of characters perform the error-detection checks. Although the relative redundancy in block-checking codes is much smaller than in the bit check on each character, their error-control efficiency is far greater. In other words, the amount of redundancy inserted into the system is by no means a guarantee of efficient error-free transmission.

Cyclic redundancy checking (which will be described later) is one of the most powerful and efficient error-detection methods in general use today. For sophisticated computer data communication, these codes are suggested for use in conjunction with retransmission. In order, however, to reduce the number of retransmission requests, some error correction should also be introduced, namely, a hybrid scheme consisting of a simple FEC system for random errors operating within an ARQ system for burst of errors. For this scheme the receiving node will request retransmission only when the number of errors is too large to be corrected locally.

12.2 ERROR-CONTROL PARAMETERS

In the previous section, a few suggestions were presented for the selection of a suitable error-control system. In this section, a description is given of the parameters for defining the required techniques and for appraising the various systems. The designer must first clarify the type of system involved, i.e. whether it is an ARQ or FEC, and then define the variables of the system. The next step is to consider the inherent reliability and efficiency of the various error-control systems and weigh them against the implementation cost.

The basic system parameters used to describe error control are as follows: reliability, efficiency, redundancy and throughput. The basic variables in the transmission system are k, which represents the number of information digits and n, which represents the total number of transmitted digits. The code is then said to be (n,k) where the number of check digits is generally $(n-k)$. The actual check digits may be smaller than $n-k$, since there may be digits in the message other than information and check digits, such as the synchronization digits, but these are ignored, as they are not required for the calculation.

Reliability is defined as the ratio of the number of correct bits to the total number of bits transmitted. Another related parameter is the coding quality, which is defined as the ratio of the number of errors after decoding to the number of prior errors.

The code efficiency is defined as the ratio of the number of information bits in the code to the total number of bits transmitted. It should be observed that this parameter may be misleading, since it does not define the actual error-control efficiency; for this reason many references have labelled it the "information rate", and this is the one suggested here. The parameter information rate is important for calculating the cost effectiveness of the transmission system:

$$\text{Information rate (Code efficiency)} = k/n.$$

Redundancy is defined as the ratio of the check digits to the total number of bits transmitted:

$$\text{Redundancy} = (n-k)/n = 1 - \text{Information rate}.$$

Throughput is defined as the ratio of the average of information digits accepted by the receiving node per unit of time to the digit rate per second of the modem. The following equation of the throughput for block codes has been suggested by Balkovic and Muench (ref. 9):

$$\text{Throughput} = n\,[1 - P(n)]\,/(n + CV).$$

where $P(n)$ is the block error probability, C is the time delay from end of transmission of one block to the beginning of the next block and V is the signalling rate of the modem in bits per second. The above throughput equation should include more delay factors, such as synchronization and buffering delays between coder and encoder.

There are two basic error-control techniques that call for attention, those that use convolutional codes and those that use block codes. For applications that require constant data stream, convolutional codes are required. In these codes the information digits and checking digits are interleaved in a continuous stream. Block codes are usually of fixed size, i.e. fixed number of digits, where each block is transmitted and checked separately.

Convolutional codes are specified by their information rate (i.e. their efficiency) which, as stated above, is the ratio of the information to the total digits transmitted. Many convolutional techniques (such as the Hagelbarger technique, which will be described later) operate with a code efficiency of 50%. That is, the information digits are equal to the number of check digits, where in the data stream a check bit is inserted after every information digit. Another parameter used for defining convolutional codes is the constraint length, which refers to the number of bits in the decoded shift register. In general, the code may correct bursts of errors up to a length which is less than the constraint length. The decoding system must then require a clean message length between bursts of errors equal to at least three times the constraint length.

Most practical error-control systems use block codes where the information digits are first constructed into blocks containing a fixed number of bits, and then the block is transmitted in a single operation, a bit at a time. In a block having n binary digits, the number of different bit combinations can be a maximum of 2^n. Of these combinations, some may be informative combinations while others may be code words M. The efficiency of the code in presenting significant error control is then $2^n \gg M$.

A basic error parameter is the "distance of the code" d (also known as "Hamming distance") which is defined as the maximum distance that exists between any two M code words within a code. The distance between two codes can be seen from the illustration given in Fig. 12.4, which demonstrates the fact that the distance for any correcting codes

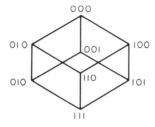

Fig. 12.4. Defining the distance of a code 'd'.

must be greater than 1. The number of errors that can be detected in any combination is $(d-1)$ where d is equal to an even number of digits and $(d-2)$ where d is an odd number of digits. In general, the number of correctable errors is equal to or less than $(d-1)/2$ for an even number and $(d-2)/2$ for an odd number. For example, if the distance is $d=2$, as in most single parity codes, then there is no possibility of recognizing double errors in the information word.

The parameter of block size is an important factor for designing an error-control system. The block size may be defined as the volume of data transmissible as a block, or it may be defined as the volume of data which must be retransmitted in the event that an error is detected. It is interesting to note that better protection is achieved with larger size blocks than with smaller ones, despite the fact that the redundancy protection is less. On the other hand, increasing the block size is clearly expensive and is inefficient with respect to retransmission and line protection. Small blocks make it possible to reduce the holding time for very short messages, although the retransmission signalling for small and large blocks is equal. As the block size increases, so does the likelihood of errors. Furthermore, block codes suffer from the disadvantage that retransmission may be required for a single error as well as for a burst.

An important parameter which defines the block size is how much retransmission is necessary. Although theoretically only the location of the error needs to be retransmitted, the hardware involved for retransmission of a single word within the block, both at the transmitter and the receiver, could be quite considerable. Furthermore, there is always a danger that the request for the retransmission of a particular location may also be distorted, and this may result in the retransmission of the wrong location. The most efficient scheme, and one which is most widely used, is to request the retransmission of the whole block. The request code for retransmission must be short, single (with no heading or address) and distinguishable from the ACK code. Although retransmission of the full block may present delays, it is still the cheapest and the most practical scheme. Block sizes of 256, 512, 960, 3860 bits are being commonly used, although the optimum is in the range of 500-1000 bits.

Another factor that must be defined is how many retransmission operations should be allowed. Any retransmission reduces the efficiency of the data-transmission system, and consequently the number must be kept to a minimum. Two requests for retransmission of the same block signify that the line is very noisy, and alternative routing should be checked. Three retransmissions, however, are more practical and are used by most systems, though in some systems even five or more retransmissions are used. The allowance of three retransmissions is the optimum figure, providing as it does some means of error correction.

12.3 BLOCK LENGTH ERROR DETECTION

A brief description of simple techniques is obviously needed before proceeding with sophisticated schemes. The most simple error-detection form in use is the checking of the received message length with that which was transmitted, to see whether the noise has caused the addition or cancellation of a digit in the sequence. By this procedure, only the length is detected while the message itself may remain in error. This technique together with simple parity checking could be efficient, but only for plain applications such as telemetry, where random messages, all of fixed length, are transmitted.

In some asynchronous data-transmission systems the clock timing reference which locks the receiver with the transmitter is transmitted with the data. This enables extra pulses to be introduced by the noise in the line and then received as clock or information digits. The basic procedure is to check whether an extra digit was introduced or whether one of the digits was eliminated during the transmission process. A simple telemetry transmitter–receiver which demonstrates this technique is shown in Fig. 12.5. The

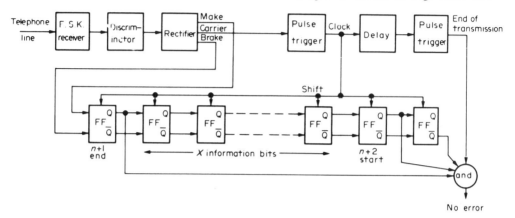

Fig. 12.5. Detection of message length deformation.

data-transmission media here is an FSK scheme of three levels, where the carrier is used to reconstruct the clock pulses. The first and last digit of each message transmitted is always a '1', and so if the information message consists of n bits, the transmitted message consists of $n + 2$ bits. The receiver's shift register consists of $n + 3$ stages. If the message is received with the correct number of digits, then the output 'AND' gate will be opened with the end of transmission pulse, that is, the sequence received is $1 \ldots 10$. If an extra pulse was picked up *en route* it will be recognized as a digit and may shift the register one extra position, thus presenting the sequence $1 \ldots \Phi1$. If a digit was lost during transmission, the sequence received may represent either $1 \ldots 00$ or $0 \ldots 10$. The efficiency of this scheme may be increased if the parity checking results are also used to operate the AND gate.

An analogue to this message-length scheme is used in more complicated schemes, although the likelihood of the addition of an extra pulse in a synchronous system is remote.

In large variable length messages the same principle is used as shown in Fig. 12.6. However, instead of transmitting '1' bits at the beginning and end of a message, special

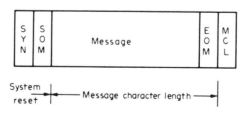

Fig. 12.6. Defining message character length.

start and end characters are transmitted. If the number of any information combination between these two characters is different from the one expected, it is immediately recognized. Where the message length is variable, an extra character may be added, labelled MCL, to represent the number of characters included in the message combination length.

Another group of codes, known as *M*-out-of-*N* codes, could also be included in the category of block length error control. In these codes the total number of digits in each character is fixed and equal to *N*, and the total number of '1's in each combination is equal to *M*. If the number of '1's received is different from *M*, i.e. whenever they are more or less than *M*, an error is recognized. These codes are also known as the "forbidden-combination check" and present simple rules for detecting forbidden combinations. The *M*-out-of-*N* codes are widely used, but as they are always restricted in detecting power to simple errors and fail on double errors, they are generally used in conjunction with other codes. The simplest code used is the 2-out-of-5 code which is applied in some computers; here, for example, 0 is represented as 00011 and 5 as 01100. This code only provides 10 combinations, which are not enough for the signalling codes for data transmission. The most widely used *M*-out-of-*N* codes are the 3-out-of-7 and 4-out-of-8.

12.4 MODULO 2 (MOD-2) ARITHMETIC

Since many sophisticated error detecting and correcting codes apply modulo 2 arithmetic, a brief description of it is presented here to facilitate the understanding of the various error-correcting codes. Modulo 2 arithmetic is very similar to the conventional binary arithmetic, except that for addition there are no 'carry' digits and for subtraction there is no 'borrow' digit. As in binary arithmetic there are only two symbols used in Mod-2 arithmetic, '1' and '0'. Mod-2 multiplication and addition are both associative operations, as in ordinary binary arithmetic. Addition (mod-2) is performed by the equations of $0 + 0 = 0, 0 + 1 = 1, 1 + 0 = 1$, and $1 + 1 = 0$. These equations are the same as in binary addition except that in the case of $1 + 1 = 0$ there is no carry ahead to the higher power of 2. Multiplication (mod-2) is performed by the same equations as in binary multiplications, that is, $0 \times 0 = 0, 0 \times 1 = 0, 1 \times 0 = 0$ and $1 \times 1 = 1$. In other words, the product is equal to 1 only if the two input digits were also 1.

Figure 12.7 illustrates the truth table for both modulo 2 addition and multiplication. It can be seen that modulo 2 addition is essentially the same as the exclusive-OR function that can be mathematically expressed as $A \oplus B$, and multiplication is the same as the AND function that can be expressed as $A \times B$. A curious property of the exclusive-OR

Fig. 12.7. Truth tables of modulo-2 arithmetic.

function is that any number added to itself will cancel out, that is, it equals zero, viz. $1 \oplus 1 = 0$ and $0 \oplus 0 = 0$. Moreover, this property leads to another interesting feature, that modulo 2 addition and subtraction are indistinguishable, that is, $A \oplus B = A \ominus B$. Thus subtraction is characterized by $0 - 0 = 0$, $0 - 1 = 1$, $1 - 0 = 1$, and $1 - 1 = 0$. This is obvious, since there are no 'carry' or 'borrow' in modulo 2. This leads to another feature, where the result of any mod-2 operation may be equal to one of the products, viz. $C = A \oplus B$, $A = B \oplus C$ and $B = C \oplus A$.

The following example demonstrates the addition (or the subtraction) of two numbers in mod-2 which omits the use of carry or borrow:

$$\text{add or subtract} \quad \begin{array}{c} 1\ 0\ 1\ 1\ 0\ 1\ 0 \\ 0\ 1\ 0\ 1\ 1\ 1\ 0 \\ \hline 1\ 1\ 1\ 0\ 1\ 0\ 0 \end{array}$$

If a whole set of numbers are involved, viz. $A \oplus B \oplus C$, the result may be reached immediately and there is no need to add each two individual numbers, as the result will always be the excess of order power of 2. For example, $1 \oplus 0 \oplus 1 = 0(2)$, that is $[1 \oplus 0] \oplus 1 = 1 \oplus 1 = 0$, or $1 \oplus 1 \oplus 1 = 1(3)$, that is, $[1 \oplus 1] \oplus 1 = 0 \oplus 1 = 1$.

Multiplication is performed in the same manner as in binary arithmetic, except that when adding the partial product no carry is considered.

$$\begin{array}{lr} \text{multiplicand} & 1\ 0\ 1\ 1\ 0\ 1\ 0 \\ \text{multiplier} & 1\ 1\ 0\ 1 \\ \hline & 1\ 0\ 1\ 1\ 0\ 1\ 0 \\ \text{add} & 1\ 0\ 1\ 1\ 0\ 1\ 0 \\ & 1\ 0\ 1\ 1\ 0\ 1\ 0 \\ \hline & 1\ 1\ 1\ 1\ 1\ 0\ 0\ 0\ 1\ 0 \end{array}$$

Division is also a simple operation, since all that is required is to add instead of subtract. In mod-2 there are no fractions, and so the division process is terminated when the number of digits in the remainder has fewer digits than the divisor. The divisor is subtracted from or added to the dividend, and for each such successful operation a 1 is recorded in the quotient. A successful subtraction operation can only be effected if the highest order of the partial product is equal to 1 and the number of digits is equal to that of the division less one. If the partial product has fewer digits than the divisor, an '0' is recorded in the quotient. This process is demonstrated in the following example:

$$\begin{array}{r} 1\ 1\ 0\ 0\ 1 \quad \text{quotient} \\ \text{Divisor } 1101\ \overline{)1\ 0\ 1\ 1\ 0\ 1\ 1\ 1} \\ 1\ 1\ 0\ 1 \\ \hline 1\ 1\ 0\ 0 \\ 1\ 1\ 0\ 1 \\ \hline 1\ 1\ 1\ 1 \\ 1\ 1\ 0\ 1 \\ \hline 1\ 0 \quad \text{remainder} \end{array}$$

The division could easily be checked by multiplying the quotient by the divisor and then adding the remainder, as shown below.

$$
\begin{array}{lrl}
\text{multiply} & 1\ 1\ 0\ 0\ 1 & \text{quotient} \\
 & \underline{1\ 1\ 0\ 1} & \text{divisor} \\[4pt]
 & 1\ 1\ 0\ 0\ 1 & \\
\text{add} & 1\ 1\ 0\ 0\ 1 & \\
 & \underline{1\ 1\ 0\ 0\ 1} & \\[4pt]
 & 1\ 0\ 1\ 1\ 0\ 1\ 0\ 1 & \\
\text{add} & \underline{\hspace{24pt}1\ 0} & \text{remainder} \\[4pt]
 & 1\ 0\ 1\ 1\ 0\ 1\ 1\ 1 &
\end{array}
$$

It is convenient in data transmission to represent the series of binary digits as the coefficients of a polynomial in the dummy variable of X. If the data series consists of n bits, then the polynomial will be of the order of $(n-1)$.

$$F(x) = a_{n-1}X^{n-1} + a_{n-2}X^{n-2} + \ldots + a_iX^i + \ldots + a_2X^2 + a_1X^1 + a_0$$

where $n > i \geqslant 0$ and a_i may be either '0' or '1' only. The polynomial is generally written high order to low order because the polynomial represents serial data. When used in this way, the '1' represents the integer of X while the '0' is represented by its omission. Hence, for example:

$$1100101 = 1 \cdot X^6 + 1 \cdot X^5 + 0 \cdot X^4 + 0 \cdot X^3 + 1 \cdot X^2 + 0 \cdot X^1 + 1 \cdot X^0 =$$

$$X^6 + X^5 + X^2 + 1$$

The advantage of the polynomial form is that the position of the high orders representing a '1' are instantly recognizable. Furthermore, modulo 2 arithmetic operations using the polynomial form are essentially the same as before, if not simpler. The addition operation is performed as before, although suitable spacing should be employed. The examples presented below are the same examples used before:

$$
\begin{array}{lccccc}
 & X^6 & + X^4 + X^3 & & + X^1 \\
\text{add} & & X^5 & + X^3 + X^2 + X^1 \\
\hline
 & X^6 + X^5 + X^4 & & + X^2
\end{array}
$$

The multiplication process is performed by multiplying the appropriate powers of the multiplier with that of the multiplicand:

$$
\begin{array}{llll}
\text{multiply} & X^6 \quad + X^4 + X^3 \quad\quad + X^1 & \text{multiplicand} \\
 & \underline{\hspace{40pt} X^3 + X^2 \quad\quad +1} & \text{multiplier} \\[4pt]
 & X^6 \quad + X^4 + X^3 \quad + X^1 & \\
\text{add} \quad X^8 \quad + X^6 + X^5 \quad\quad + X^3 & \\
\underline{X^9 + \quad + X^7 + X^6 \quad\quad + X^4} & \\[4pt]
X^9 + X^8 + X^7 + X^6 + X^5 \quad\quad\quad\quad\quad + X^1
\end{array}
$$

In the division operation, the divisor must be brought to the order level of the dividend before it is subtracted (or added) from the dividend. The highest order level of the quotient is obtained by subtracting the highest order of the divisor from that of the dividend:

$$
\begin{array}{r}
X^4 + X^3 \qquad\qquad + 1 \quad \text{quotient}\\
X^3 + X^2 + 1 \overline{)\, X^7 \qquad + X^5 + X^4 \qquad + X^2 + X^1 + 1}\\
X^7 + X^6 \qquad + X^4 \qquad\qquad\qquad\\
\overline{\qquad X^6 + X^5 \qquad\qquad\qquad\qquad}\\
X^6 + X^5 \qquad + X^3 \qquad\qquad\\
\overline{\qquad\qquad X^3 + X^2 + X^1 + 1}\\
X^3 \;+ X^2 \qquad + 1\\
\overline{\qquad\qquad\qquad X^1 \quad \text{remainder}}
\end{array}
$$

As before the division operator may be checked by the operation of

$$(X^4 + X^3 + 1) \cdot (X^3 + X^2 + 1) \oplus (X^1).$$

12.5 PARITY ERROR-CONTROL CODES

Parity codes are widely used in simple data-transmission systems, their implementation being relatively inexpensive. This code is used in two common forms, one on a character check basis and the other on a block basis. Both forms are based on the concept of parity which represents the modulo 2 addition of all the digits in a character or block.

In this method the resultant digit, known as the parity digit, is transmitted at the end of the character, that is, in a simple word with n digits, of which one digit is used for error detection and the other digits equal to $k = (n-1)$ are information digits. The value of the parity is so chosen that it is equal to 1 when the total number of '1's in the character is odd. As explained before, the process to obtain the parity digit is by mod-2 adding all the digits. If the character contains a sequence of k information digits, then the parity will take the following form:

$$P = a_k \oplus a_{k-1} \oplus \ldots \oplus a_i \oplus \ldots \oplus a_2 \oplus a_1 \oplus a_0$$

where a_i is either '0' or '1'. For example, if the information character is 1101, the parity digit will be equal to 1, since there is an odd number of '1's ($1 \oplus 1 \oplus 0 \oplus 1 = 1$) or if the information character is 0110, the parity digit will be equal to 0. The circuit for

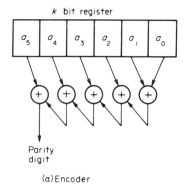

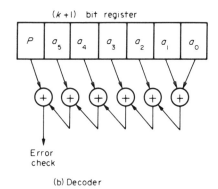

Fig. 12.8. Parity coding circuitry.

generating this equation is simple, as can be seen from the illustration given in Fig. 12.8 (a). The parity equation could also be written as follows:

$$0 = P \oplus a_k \oplus a_{k-1} \oplus \ldots \oplus a_i \oplus \ldots \oplus a_2 \oplus a_1 \oplus a_0.$$

From the above equation it is easy to design the decoding circuitry for detecting an error, as seen from Fig. 12.8 (b). An extra stage is added to the receiver register for the parity check which was transmitted after the information character. The output of the circuit gives an '0' if no error is detected and a '1' if an error is detected. That is, it gives a positive indication when an error occurs.

Similarly, there are even parity check codes which check whether the total number of '1's in the character is even. The equations for the even parity will take the form of

$$P = a_k \oplus a_{k-1} \oplus \ldots \oplus a_i \oplus \ldots \oplus a_2 \oplus a_1 \oplus a_0 \oplus 1,$$

$$1 = P \oplus a_k \oplus a_{k-1} \oplus \ldots \oplus a_i \oplus \ldots \oplus a_2 \oplus a_1 \oplus a_0.$$

In some publications this code is also referred to as "odd parity", where they include the parity digit when summing the '1's. There are no arguments needed in favour of either odd or even parity, as they both produce the same results and both are widely used. However, it should be added that with even parity a character of all zeros is illegal.

The parity checking can detect the existence of any errors in the character provided that an odd number of errors has occurred, including those in the parity itself. However, it cannot detect an error if there was an even number of errors, and it cannot locate the position of the error.

The circuitry shown in Fig. 12.8 determines the parity digit by means of a parallel operation. Similarly, the parity may be determined by serially complementing a T-type flip-flop which is initially set to the '0' state for an odd parity or to the '1' state for an even parity.

There are several methods using parity error detection for block transmission, two of which are mentioned here. The first is the technique known as Longitudinal Redundancy Checking (LRC) and is one which basically uses the simple parity technique, except that all the parity digits are transmitted as a separate character at the end of the block.

Separating the check digits from the information digits has the advantage that it is less likely to have distortions in both information and check regions of the block. The checking code is also referred to as row checking or horizontal checking, as the block is usually in rectangular array form. In this technique the parity of successive information characters in the block is added to the contents of an LRC register. After the last character of the block has been checked, the resultant which is contained in the LRC register is transmitted immediately following the block. The LRC technique will detect all error patterns involving an odd number of errors within each character.

In the second technique, the longitudinal redundancy checking which is used concurrently with a vertical redundancy check is usually referred to as LRC/VRC. In the LRC technique, the block is checked horizontally, while in the LRC/VRL coding technique the block is checked both horizontally and vertically for parity digits. In the vertical checking, the digits in the column are checked, i.e., digits of the same relative order of different characters. As in the LRC technique, each bit of the column check is placed in the VRC register. The last digit in the VRC register (i.e. the corner of the block) is an LRC check on the VRC register, although it could also be a VRC check on the LRC register. Both the LRC and the VRC register contents are transmitted after the transmission of the information block.

The VRC/LRC coding technique is also called 'geometrical' codes and "block parity" codes since it is made up of a rectangular array of V characters, each of L digits, as shown in Fig. 12.9. The block size M is equal to $(V + 1) \cdot (L + 1)$, while $(L \cdot V)$ are the information

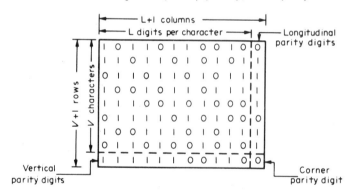

Fig. 12.9. Block (geometric) parity check.

digits and $(L + V + 1)$ are the checking digits. The minimum code distance of this technique is L. The redundancy of the code is equal to

$$\frac{L+V+1}{(V+1)\,(L+1)}$$

If the block consists of 14 characters, each of 7 bits, then the redundancy is equal to 22/120, which is about one-sixth as many check digits as information digits. With larger blocks there is less redundancy, although the level is still high.

This code can detect an odd number of errors within each character or within each bit position of all the characters of the block. It may also detect even numbers of errors in the block, provided they do not occur in the same intersection of the row and column. Furthermore, this coding technique provides some error correction, as it can detect the location of some of the errors. The detection of the location follows the following rules:

(a) An error in the information may be located at the intersection of a row and column, provided only one row and one column check have failed.

(b) An error in a row parity digit may be located if only one row digit has failed but with no column failure.

(c) An error in a column parity digit may be located if only one column digit has failed, with no row failure but with the corner failure.

(d) An error in the corner parity digit may be located if only the corner digit has failed, with no column or row digit failure.

The block parity code is commonly used for simple applications, as it may provide good error-burst detection capabilities and single digit error correction. It may not, however, locate all the errors, since double errors may escape through the decoder checking.

12.6 HAMMING ERROR-CORRECTING CODES

The simple character parity check may only detect a single error and cannot locate the error digit position within the character; hence the error cannot be corrected. Hamming has developed a technique by which a single error could be corrected. This is achieved by increasing the number of check digits.

The Hamming single-error code is formulated by first determining how many check digits are needed. In principle, an error-detection code requires a minimum distance of two to detect an error, and an error-correcting code requires the use of the minimum distance of three. If there are only one information digit and one parity check digit, an error will be detected if the result does not tally. However, it is not enough to detect the location, since the mistake may be either in the information digit or in the check digit. If the message consists of one information digit labelled M, then two parity digits labelled P_1 and P_2 must be added. The three digits transmitted can generate the error location by two check operations, viz. C_1 and C_2:

$$\text{check } C_1 \quad 0 = P_1 \oplus M,$$

$$\text{check } C_2 \quad 0 = P_2 \oplus M.$$

An error is located in the information digits if both check tests have failed, but if only one test has failed, then the error is in one of the two parity digits. If the three digits are placed in a register in the order P_1 in the first position, P_2 in the second position and M in the third position, then the binary combination of the result of the two check tests will indicate in which position the error occurred (Table 12.1).

With larger messages more parity digits are required, although the redundancy is reduced. If the number of digits transmitted is n and the number of information digits k, then the number of parity check digits is $n-k$. With $n-k$ check digits there are $2^{(n-k)}$ possible combinations which could be detected. As one of these must represent the case of no errors, single errors could be detected in the string of $2^{(n-k)} - 1$. Thus, for a given

TABLE 12.1.

C_2	C_1	test check
0	0	no error
0	1	error in position 1 (P_1)
1	0	error in position 2 (P_2)
1	1	error in position 3 (M)

number of information digits k, the number of check digits could be calculated by the following equation:

$$2^{(n-k)} - 1 \geqslant n.$$

As k and n must not be a fraction, the result is taken to the lowest nearest integer. For a given total of n bits, the nearest integer has been reported by Hamming as shown in Table 12.2.

TABLE 12.2

n	k	n-k
3	1	2
4	1	2
5	2	3
6	3	3
7	4	3
8	4	4
9	5	4
10	6	4
11	7	4
12	8	4

Let us consider, for example, the conventional code $k = 4$ information digits which require 3 check digits, giving a message with the total of $n = 7$ digits. Two questions arise in designing the code, where to insert the 3 parity digits and what are the required parity check tests. Since $n = 7$ there will be 7 positions, and the test checks must point to each of them. Table 12.3 indicates which tests are required, using the binary notation to indicate each position.

From the Table 12.3 test, C_1 will detect an error by the combination of the data in positions 1, 3, 5, 7, test C_2 in positions 2, 3, 6, 7 and test C_3 in positions 4, 5, 6, 7. The parity is selected as the most significant digit of the position number of each test. In other words, the parity for test C_1 will be in position 1, for test C_2 in position 2 and for test C_3 in position 4. The parity check corresponds to the position of 2^i where i is equal to $0 < i < (n-k-1)$. Therefore, the information digits will be in position 3, 5, 6, 7 and the parity digits in positions 1, 2 and 4. This gives the following three check positions which may be obtained by the circuit shown in Fig. 12.10:

C_1 check $\quad P_1 = M_0 \oplus M_1 \oplus M_3,$

C_2 check $\quad P_2 = M_0 \oplus M_2 \oplus M_3,$

C_3 check $\quad P_3 = M_1 \oplus M_2 \oplus M_3.$

TABLE 12.3

C_3	C_2	C_1	test check
0	0	0	no error
0	0	1	error in position 1
0	1	0	error in position 2
0	1	1	error in position 3
1	0	0	error in position 4
1	0	1	error in position 5
1	1	0	error in position 6
1	1	1	error in position 7

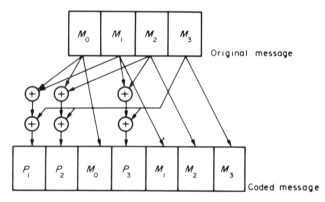

Fig. 12.10. Generating Hamming code with conventional logic.

TABLE 12.4

	Position	1	2	3	4	5	6	7	
	Digit	P_1	P_2	M_0	P_3	M_1	M_2	M_3	
Information data				0		0	1	1	
P_1 parity digit		1		0		0		1	
P_2 parity digit			0	0			1	1	
P_3 parity digit				0		0	1	1	
Transmitted message		1	0	0	0	0	1	1	
Received message		1	0	1	0	0	1	1	
C_3 check 4, 5, 6, 7					0	0	1	1	$C_3 = 0$
C_2 check 2, 3, 6, 7			0	1			1	1	$C_2 = 1$
C_1 check 1, 3, 5, 7		1		1		0		1	$C_1 = 1$

The example in Table 12.4 demonstrates the operation of the Hamming code when a message of 0011 was transmitted and the message of 1011 was received.

When three check results give $C_3 \, C_2 \, C_1 = 011$, the error is in position 3 (i.e. M_0) and this can now not only be detected but also corrected.

By adding a fourth parity digit the code will be $(8, 4)$ and the distance will be 4 instead of 3. This fourth parity check digit will allow double error detection, although still only having single error-correction capability. With a distance of 5 it is possible to obtain double error correction, and with a distance of 7 a triple error correction is possible. Theoretically, increasing the distance will increase the error-control capabilities; however, the physical implementation of the combinatory form is complicated. A more attractive approach to increase the error-control capabilities is through the use of cyclic codes.

12.7 CONVOLUTIONAL ERROR-CONTROL CODES

The parity-type codes described so far are intended to provide error control for fixed-size format information. The codes described in this section are for continuous processing of data with check digits interspersed among the information digits. These codes are known both as Convolutional codes and Recurrent codes. The codes, though not of a fixed blocked structure, are still also defined as an (n, k) code in the same manner as in block codes. For each group of n digits transmitted, the k are information digits and $(n-k)$ are checking digits. Convolutional codes are specified by their code rate (k/n) and their constraint length N. The n-digit code group depends not only on the k-digit information group of the same unit but also on the previous $(N-1)$ message groups. For convolutional codes, the parameters k and n are generally small integers. Thus, typical code rates are of $1/2$ which use $(2, 1)$ presenting a stream of data digits having a parity check digit computed for each information digit. There are possibly other convolutional code rates which present better efficiency, but these cannot be discussed here.

The convolutional codes are simple in structure, easily understood and easily mechanized. They are intended to detect and correct error bursts of up to a length of $2m$ digits in a sequence. It must be noted that once an error burst of the full m length has occurred in a stream it must be followed by an error-free space of $(6m + 1)$ digits before it can correct errors again.

One of the early classical convolutional error-correcting codes is that suggested by Hagelburger. The advantage of that technique is that the information digits flow continuously with no need to split the data into blocks. The encoder, shown in Fig. 12.11 (a), consists of a shift register of the length of $(2m + 1)$ stages and a modulo-2 adder connected between two stages of the register. In the example shown, each information digit is compared with the fifth digit in the sequence:

$$P = M_0 \oplus M_4.$$

The check parity digits are transmitted between each two information digits, but, to avoid the possibility of an error in both the adjacent parity and information digits, their transmission is relatively delayed. That is, the information digits, once they have been

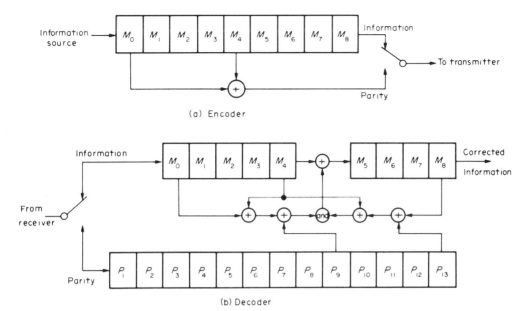

(a) Encoder

(b) Decoder

Fig. 12.11. Hagelbarger error-correcting technique.

checked, are delayed in the register for a given length after the parity was transmitted. If the parity is between digits 1 and $(m + 1)$, the delay will be for the length of m, and thus the shift register will be the length of $(2m + 1)$. The coding process may be better understood with the example given in Fig. 12.12, which shows the coding of the sequences of information digits and the transmitted stream.

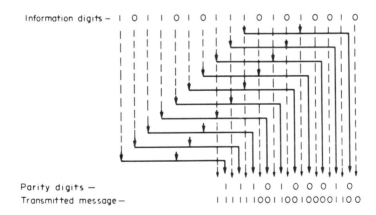

Fig. 12.12. Encoding chart of the Hagelbarger technique.

At the receiver, the stream of digits must be split into two separate streams, one of information digits and the other of parity digits, which are inserted into two separate shift registers, as shown in Fig. 12.11 (b). The operation of the decoder is similar to that of the encoder, namely, comparing the two digits, M_0 and M_4, in the information stream with the associated parity digit. Since the parity digit was transmitted ahead of the information, it must now be delayed for the length of $(2m + 1)$ before the associated information digits arrive. If the result of the comparison is equal to 0, there is no error, but if it equals 1 this means there is an error.

$$P_9 \oplus M_0 \oplus M_4 = 0 \quad \text{no error,}$$

$$P_9 \oplus M_0 \oplus M_4 = 1 \quad \text{error.}$$

However, this result does not give enough information about the location of the error, which could be in any one of the two information digits or in the parity digits. To remedy this defect, both the information and the parity digits are delayed for another m stage. This will enable the M_4 information digit to be compared twice, once with the corresponding previous digit M_8 and once with the corresponding following digit M_0. Only if both comparison checks show an error can it be assumed that M_4 is in error.

$$P_9 \oplus M_0 \oplus M_4 = 1 \qquad P_{13} \oplus M_8 \oplus M_4 = 1.$$

Where such an error is located, the information in M_4 is complemented as it is shifted into M_5 stage. The complementing operation may be enforced by a modulo 2 process. An error is recorded when a 1 is received from the binary AND gate, indicating that both checking operations have failed. If M_4 has information 0, it will be complemented to 1 by $0 \oplus 1 = 1$, and vice versa if it has information 1, $1 \oplus 1 = 0$.

If there is a double error in the information digits and in the corresponding parity digits, then the error will pass undetected. In fact, this technique can correct errors up to bursts of $(2m + 1)$ but not beyond. Once such an error has occurred there must be a free no-error gap of $(6m + 1)$ before the encoder can again work properly. Another disadvantage of this technique is the delays incorporated into the system.

An improvement on the Hagelberger scheme, shown in Fig. 12.13, requires less

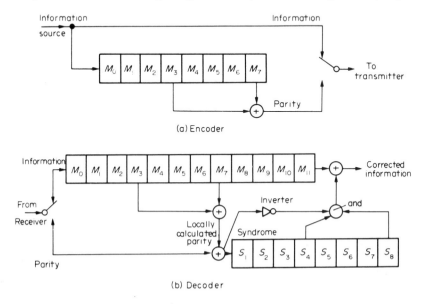

Fig. 12.13. Convolutional error correction with syndrome calculation.

equipment and results in less overall delays. The encoder consists of a shift register of $2\,m$ stages only. The information digits are transmitted with no delay, while the parity check digits are delayed by the shift register. As before, two sequential streams of digits are obtained, which are alternately intercoded, with one information digit followed by one parity digit.

At the receiver the received digits are again divided into two streams, information and parity. As the information stream is in fact a copy of the real one transmitted, the parity digits may be computed again at the receiver and the results may then be compared with the parity digits received. Furthermore, the information digits must now be delayed so as to correspond with the associated parity digit. The parity comparison of the stream received with the newly computed stream is done by modulo 2 addition. The output of this addition is called 'syndrome'. If the two inputs are equal, implying that no error is detected, the syndrome will be equal to zero. However, if they are not equal, thus indicating an error, the syndrome will equal to a one. The syndrome results are then placed in a shift register which now consists of error patterns in both the information and/or the parity digits. From this it is obvious that there is no need to check the information digits twice as in the Hagelberger technique, since the error information will now be memorized by the syndrome. In other words, the syndrome inserted into the register is equal to

$$S = P \oplus M_3 \oplus M_7.$$

This shows that the result of the comparison of M_{11} with the previous relative information digit is retained in S_8 and the result of comparing it with the following information digit is retained in S_4. An error can be then detected and corrected if the error patterns in both S_8 and S_4 tally. To increase the reliability of the check, S_4 is compared with the new syndrome, to ensure that this error information does not refer to M_7. In this way, if there is an error indication in the new syndrome, it will inhibit the output of the AND gate of S_4 and S_8. This last modification requires $(m-1)$ stages added to each register, although the overall number of stages is still one stage less for each register compared with the original Hagelberger technique.

There is always a danger that there will be an error in both the parity and information digits despite the relative delay between them. An alternative approach is to code the parity digit according to a more restricted code pattern but to relax the checking process. An error may then be detected and corrected by majority rule, or by a given threshold, rather than by fixed mathematical results. This convolutional error coding is shown in Fig. 12.14, which demonstrates how the parity can be coded by a more restricted pattern.

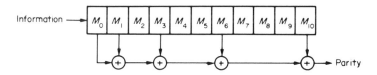

Fig. 12.14. Parity generator for $P = M_0 \oplus M_1 \oplus M_3 \oplus M_6 \oplus M_{10}$.

The information digit is compared not with one other following digit but with a number of them. The comparison chosen is $m = 10$, where M_{10} is compared with digits spaced apart by 1, 2, 3 and 4 stages.

$$P = M_{10} \oplus M_6 \oplus M_3 \oplus M_1 \oplus M_0.$$

The principle of the encoder and decoder is the same as shown before in Fig. 12.13, except that the shift register is of the length of m instead of $2m$. In the decoder shown in Fig. 12.15, the syndromes are calculated in the same manner as before, and so are the

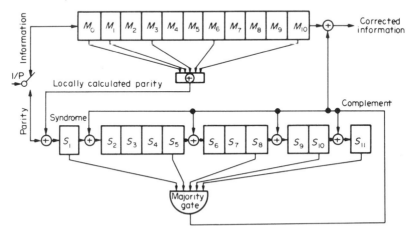

Fig. 12.15. Error correction by majority logic of syndromes.

relative outputs taken from the syndrome shift register. However, instead of taking these syndrome outputs to an AND gate, they are taken to a majority gate, and so if there were one or more errors in transmitting the code they would still be recognized. In fact, this majority gate scheme may correct up to three errors which code the same information digits. Once the error is detected and corrected, the corresponding syndrome must also be complemented so that the encoder can handle other error patterns.

An important aspect of the error-correction systems discussed here is the synchronization of the receiver with the transmitter. As there are only two possibilities for the digit orientation, either parity or information, the decision may be carried out by trial and error. That is, if there is a continuous stream of errors, this would first indicate that the system is not synchronized. It is essential that a continuous stream of '1's or of '0's will not be used, because this will cause the encoder to be stuck in one of two positions.

12.8 GENERATING CONVOLUTIONAL PARITY DIGITS

All the parity generators of convolutional codes discussed so far have the interesting property of modulo 2 multiplication of the input information stream by a fixed code. The coding circuits shown in the previous sections could be represented as polynomial $P(X)$, exhibiting the fixed code with which the input information polynomial $M(X)$ is multiplied so as to provide the coded output stream.

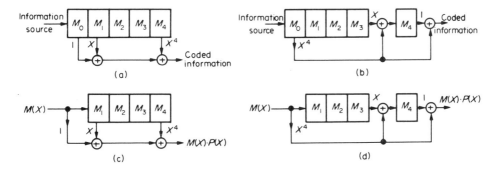

Fig. 12.16. Circuits for multiplying by polynomial $P(X) = X^4 + X + 1$.

For the purpose of explaining this process, we may use the example used in Fig. 12.16 (a). This example is drawn up on the same principle as the one before where the parity generated is equal to $P = M_0 \oplus M_1 \oplus M_4$. The same circuit could be represented by the fixed polynomial $P(X) = X^4 + X + 1$, which is known as the generator polynomial, since it is the one which multiplies the input polynomial to receive the check polynomial, viz.

check polynomial = (information polynomial) · (generator polynomial)
$$C(x) = M(x) \cdot P(x).$$

It should be pointed out that the transmitted data stream consists of the interleaving of both the check polynomial $C(X)$ and the information polynomial $M(X)$, as explained in the previous section.

The modulo 2 multiplication process may be proved by comparing the multiplication of the two polynomials with the step-by-step shifting of the information through the coder. For the example, let us suppose that the information stream of 11101 is sent through the coder and is represented by the polynomial $M(X) = X^4 + X^3 + X^2 + 1$:

$$
\begin{array}{r}
X^4 + X^3 + X^2 \qquad + 1 \; M(X) \\
\text{multiply} \qquad\qquad X^4 \qquad\qquad + X + 1 \; P(X) \\ \hline
X^4 + X^3 + X^2 \qquad + 1 \\
X^5 + X^4 + X^3 \qquad + X \\
X^8 + X^7 + X^6 + \qquad + X^4 \\ \hline
X^8 + X^7 + X^6 + X^5 + X^4 \qquad + X^2 + X + 1 \; C(X) = M(X) \cdot P(X)
\end{array}
$$

The check polynomial received by the multiplication is equivalent to ... 111110111 .... The same multiplication process is compared with the step-by-step operation shown in Table 12.5. The input stream is inserted, a digit at a time, into the shift register shown in Fig. 12.16 (a).

The circuit used for the example shown in Fig. 12.16 (a) is the standard circuit used throughout this chapter for coding convolutional codes. The same coding process could be obtained by another circuit, shown in Fig. 12.16 (b), which also conforms to the

TABLE 12.5

Input information					Shift register					Coded output	
X^4	X^3	X^2	X^1	X^0	M_0	M_1	M_2	M_3	M_4	parity	polynomial
1	1	1	0	1	0	0	0	0	0	0	
0	1	1	1	0	1	0	0	0	0	1	1
0	0	1	1	1	0	1	0	0	0	1	X
0	0	0	1	1	1	0	1	0	0	1	X^2
0	0	0	0	1	1	1	0	1	0	0	
0	0	0	0	0	1	1	1	0	1	1	X^4
0	0	0	0	0	0	1	1	1	0	1	X^5
0	0	0	0	0	0	0	1	1	1	1	X^6
0	0	0	0	0	0	0	0	1	1	1	X^7
0	0	0	0	0	0	0	0	0	1	1	X^8
0	0	0	0	0	0	0	0	0	0	0	

modulo 2 multiplication of the same generator polynomial of $P(X) = X^4 + X + 1$. To prove this fact the process will again be calculated by the step-by-step operation (Table 12.6).

TABLE 12.6

Input information					Shift register					Coded output	
X^4	X^3	X^2	X^1	X^0	M_0	M_1	M_2	M_3	M_4	parity	polynomial
1	1	1	0	1	0	0	0	0	0	0	
0	1	1	1	0	1	0	0	0	0	1	1
0	0	1	1	1	0	1	0	0	1	1	X
0	0	0	1	1	1	0	1	0	0	1	X^2
0	0	0	0	1	1	1	0	1	1	0	
0	0	0	0	0	1	1	1	0	0	1	X^4
0	0	0	0	0	0	1	1	1	1	1	X^5
0	0	0	0	0	0	0	1	1	1	1	X^6
0	0	0	0	0	0	0	0	1	1	1	X^7
0	0	0	0	0	0	0	0	0	1	1	X^8
0	0	0	0	0	0	0	0	0	0	0	

The first stage of the shift registers may be omitted without it affecting the coding process. This is because the front stage only stores the contents of the first digit of the input information polynomial as it is inserted into the shift register. In this way it may save a register stage in each coder, as seen from the circuits given in Figs. 11.16 (c) and (d) which generate the same fixed polynomial.

A scheme for error detection could be achieved by simply decoding the received message by dividing it with the same generator polynomial as used in the transmitter. This is possible only if the generator polynomial used for multiplication and division is one which exhibits the property of maximum length, as will be shown in the following section. If the division produces a zero remainder, the result obtained is the information polynomial. If the division, however, produces a non-zero remainder, an error is detected.

To achieve also error-correction capabilities, more information must be transmitted. In any case, this coding and decoding process may be popular when transmitting statistical or similar information. For these applications error correction may be of little importance provided the number of errors is relatively small and all the errors could be detected and so omitted in the statistical processing.

Shift register coding, especially those with feedback, are continually used for cyclic codes, and are described in the following section.

12.9 CYCLIC CODES BASIC PROPERTIES

Cyclic codes are widely used because they present a most efficient error-detection technique with relatively little redundancy. They are also reasonably simple and inexpensive to implement. Extensive studies have been centred on the sophisticated codes which provide good error protection. Although these codes are capable of both error detection and error correction, they are most frequently used only for error detection, while the correction is dealt with by a request for retransmission. The parity-check patterns used in these codes are unique and ensure that no two errors remain undetected. These codes are intended for fixed size blocks, where the size of the block may be chosen to meet the application requirement. A cyclic code with 16 check digits is considered adequate for block sizes of 240, 480 or 960 information digits. If a two-diameter block code were used, as described before, it would require at least four times as many check digits.

The term "cyclic codes" is derived from the cyclic shift register (also known as ring counter), shown in Fig. 12.17 (a), where the information stored in the register is shifted

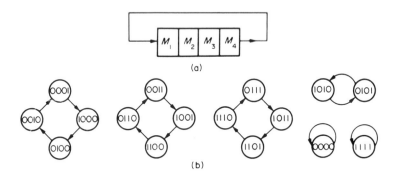

Fig. 12.17. Cyclic shift register and its state diagrams.

in a circle, with the output of the register being connected to the input. Any information pattern contained in the register will circulate and present a different pattern after each shift, provided only that the information is not all 0's or all 1's. If the number of stages in the cyclic shift register is r, then after a maximum of r shifts the same information pattern will be presented in the shift register (as shown by the state diagrams given in Fig. 12.17 (b)).

Introducing modulo 2 sum gates in the feedback path of the shift register will increase the number of possible pattern states. In the example given in Fig. 12.18, the number of

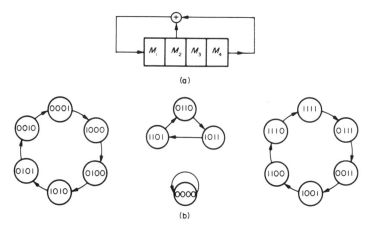

Fig. 12.18. A four-cycle feedback shift-register and its state diagrams.

states in a given cycle was increased to six. Furthermore, this circuit presents four different cycles as compared with five cycles in the ring counter. For the circuit to be useful for error coding only a maximum of two cycles may be operative, where one of the cycles is for all zeros while the other cycle contains all the possible states.

In any register of r stages there could be up to 2^r different information patterns, that is, different state patterns. The state pattern of all zero's is of no consequence in any error coding, since 0000 will always shift to the same state pattern. Hence, the maximum number of different state patterns that may be presented in each register is $2^r - 1$. An example of such a shift register with modulo 2 sum in its feedback path is given in Fig. 12.19, where the four information digits pass through $(2^r - 1) = 15$ state patterns after each shift before they return to the original state pattern. A feedback shift register which may present a $2^r - 1$ state pattern in a single cycle is known as "maximum-length shift register", since the circuit may give the maximum length possible of unique combinations. This type of shift register is the one used for cyclic codes, because it presents the maximum non-repeated code sequence which can be generated with short registers. In effect, the end-around shift of any code word produces another code word.

Feedback shift registers which possess the feature of maximum length also have the property of acting as dividers of any input sequence. Any other type feedback shift register which does not possess a single circulating cycle will not be able to perform the division operation. Moreover, if the wrong circuit is used, the information may circulate indefinitely, without ever returning to zero.

An example demonstrating a division circuit is given in Fig. 12.20, which is based on the circuit shown in Fig. 12.16 (d). There the circuit was used for multiplication purposes, while to achieve division the input and output are interchanged. To prove this operation process, an input sequence of 111110111 is used, selected to comply with the similar example given in the previous section for the multiplication of the input information. The same circuit is used except that there a feedforward was used for multiplication while here a feedback is used for division. In other words, if $P(X) =$

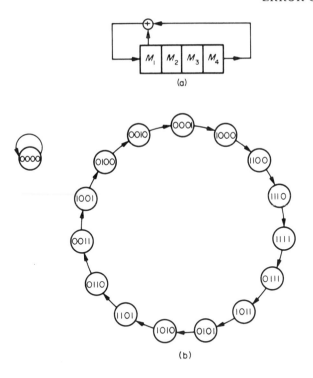

Fig. 12.19. A feedback shift-register which possesses the property of maximum length.

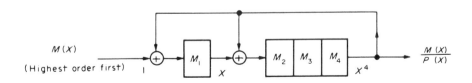

Fig. 12.20. Circuit for dividing by polynomial $P(X)=X^4 + X + 1$.

$X^4 + X^4 + 1$ was used as the generator polynomial for multiplication, then the same generator polynomial is used for the division operation. The process can be shown by the following example calculation where the information polynomial taken is

$$M(X) = X^8 + X^7 + X^6 + X^5 + X^4 + X^2 + X + 1,$$

which was the result of the multiplication obtained from the example in the previous section. The quotient of the division $M(X)/P(X)$ must present the same initial polynomial used for that example, while with multiplication the lowest order is taken first. It is essential to note that with division the highest order must be fed in first. Table 12.7 shows the step-by-step calculation of the process.

TABLE 12.7

Input information									Shift register				Coded output	
X^0	X^1	X^2	X^3	X^4	X^5	X^6	X^7	X^8	M_1	M_2	M_3	M_4	division	polynomial
1	1	1	0	1	1	1	1	1	0	0	0	0	0	
0	1	1	1	0	1	1	1	1	1	0	0	0	0	
0	0	1	1	1	0	1	1	1	1	1	0	0	0	
0	0	0	1	1	1	0	1	1	1	1	1	0	0	
0	0	0	0	1	1	1	0	1	1	1	1	1	1	X^4
0	0	0	0	0	1	1	1	0	0	0	1	1	1	X^3
0	0	0	0	0	0	1	1	1	1	1	0	1	1	X^2
0	0	0	0	0	0	0	1	1	0	0	1	0	0	
0	0	0	0	0	0	0	0	1	1	0	0	1	1	1
0	0	0	0	0	0	0	0	0	0	0	0	0	0	

The same process could be proved by the mathematical modulo 2 division:

$$
\begin{array}{r}
X^4 + X^3 + X^2 \qquad + 1 \quad \text{quotient} \\
\text{Divisor } X^4 + X + 1 \; \overline{\left)\; X^8 + X^7 + X^6 + X^5 + X^4 \qquad + X^2 + X + 1\right.} \\
\underline{X^8 \qquad\qquad + X^5 + X^4} \\
X^7 + X^6 \\
\underline{X^7 \qquad\qquad + X^4 + X^3} \\
X^6 \qquad + X^4 + X^3 + X^2 \\
\underline{X^6 \qquad\qquad\quad + X^3 + X^2} \\
X^4 \qquad\qquad + X + 1 \\
\underline{X^4 \qquad\qquad + X + 1} \\
\end{array}
$$

No remainder

A polynomial which is used for a shift register generator for maximum length sequence is known as a primitive polynomial, and therefore it divides 2^r-1 but not any smaller r.

As shown in the previous section, there could be two similar circuits performing the multiplication process. Likewise, there are two similar circuits that could perform the division process.

12.10 CYCLIC ERROR-DETECTION CODES

In the previous section it was established that the division by a primitive polynomial presents the maximum length sequence of 2^r-1 codes, which are each unique and unrepetitive. These codes of the length of r could be used to encode blocks of the size up to 2^r-r-1. Thus, by suitably selecting the generator polynomial, it is possible to create error-detecting codes of (n,k) where $n \leqslant 2^r-1$ and the number of the parity digits is equal to $r = n-k$.

In Section 12.8 it was proposed to create codes where the input message polynomial $M(X)$ is multiplied by the generator polynomial $P(X)$ in the encoder. With that method the transmitted polynomial $T(X)$ is coded and is dissimilar in reference to the original information polynomial, although the information polynomial may be recovered by the division of the received $T(X)$ by the same generator polynomial. It is far more simple and satisfactory to transmit the original information polynomial which consists of k unaltered information digits followed by $r = n-k$ check digits. Such codes are termed systematic, and can be achieved by division instead of multiplication, presenting the same error-control properties.

The transmitted polynomial $T(X)$ in cyclic codes is accomplished by the transmission of the modulo 2 sum of the input information polynomial $M(X)$ and the check polynomial $R(X)$. (It will be shown later that this check polynomial is the remainder of the division of the input polynomial by the generator polynomial.) The format of the transmitted polynomial is shown in Fig. 12.21, which also demonstrates that the highest

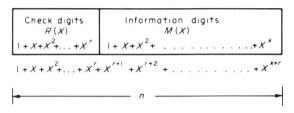

Fig. 12.21. Transmission polynomial format $T(X)=1 + X + \ldots + X^n$.

order is transmitted first. In order that both polynomials appear in sequence, first $M(X)$ then $R(X)$, and so before the addition, the input polynomial $M(X)$ must be followed by r zero states (i.e. in the lower-order positions). These r zero states will later be used for inserting the check digits. This process can be achieved by shifting $M(X)$ in the buffer register by r stages, and it can be expressed mathematically by multiplying the input polynomial by X^r. If the highest order of $M(X)$ was k the new polynomial will now be $k + r$, which is equal to the transmitted length of n. Thus, the transmitted polynomial could be expressed as follows:

$$T(X) = X^r M(X) \oplus R(X).$$

The coding is accomplished by a cyclic circuit (discussed before) performing the division of the shifted input information polynomial $X^r M(X)$ by the generator polynomial $P(X)$. This gives a unique quotient polynomial $Q(X)$ and a remainder of polynomial $R(X)$. This operation can be expressed by the following equation:

$$\frac{X^r M(X)}{P(X)} = Q(X) \oplus \frac{R(X)}{P(X)}.$$

Since modulo 2 addition and subtraction are the same, the above equation could be rewritten:

$$\frac{X^r M(X)}{P(X)} \oplus \frac{R(X)}{P(X)} = Q(X).$$

Multiplying both sides by the generator polynomial $P(X)$ will give us the required transmitted polynomial:

$$T(X) = X^r M(X) \oplus R(X) = Q(X) \cdot P(X).$$

From the above equation it can be seen that the remainder of the division $R(X)$ produces the check digits that are used in the transmitted polynomial.

Implementing the above encoding process is simple, as it utilizes the cycle circuit described in the previous section. The information sequences are shifted into the circuit with the highest order first until the last coefficient is shifted into the lowest order of the circuit. This requires a total of n shifts, k for entering the information sequence and r for shifting it to the lowest position. An example of a circuit is shown in Fig. 12.22, which is based on the generator polynomial of $P(X) = 1 + X + X^4$.

The process is better understood with the following example, where the message polynomial taken is equal to $M(X) = X^8 + X^6 + X^3 + X^2 + 1$, thus having $k = 9$. Shifting this polynomial is by 4 orders, since $r = 4$ will give $X^r M(X) = X^{12} + X^{10} + X^7 + X^6 + X^4$. Table 12.8 shows the step-by-step process where the divisor used is the one shown in Fig. 12.22.

TABLE 12.8

Input sequence		Shift register				Output sequence	
		M_1	M_2	M_3	M_4		
	X^{12} 1	0	0	0	0	0	
	0	1	0	0	0	0	
	X^{10} 1	0	1	0	0	0	Delay of r
	0	1	0	1	0	0	
Original	0	0	1	0	1	1 X^8	
message	X^7 1	1	1	1	0	0	
$M(X)$	X^6 1	1	1	1	1	1 X^6	Quotient
	0	0	0	1	1	1 X^5	$Q(X)$
	X^4 1	1	1	0	1	1 X^4	
	0	0	0	1	0	0	
Added 0's	0	0	0	0	1	1 X^2	
X^r	0	1	1	0	0	0	
	0	0	1	1	0		Remainder $R(X)$

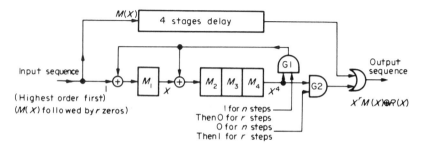

Fig. 12.22. Basic circuit for encoding block codes with the generator polynomial of $P(X) = 1 + X + X^4$.

Thus, after total of n shifts the remainder is in the register, and to receive it another r shifts are required, and so a total of $r + n$ shifts is required to get the output polynomial of the remainder. The transmitted polynomial $T(X)$, as shown in Fig. 12.21, must consist of the information polynomial $M(X)$ followed by the check polynomial $R(X)$. However, since the input information sequence is of only k digits, there is a delay of $n-k = r$ between the two polynomials, and thus the information polynomial must be delayed $n-k = r$ stages so that when it is added to the remainder they will be in a consecutive sequence.

Returning to the circuit shown in Fig. 12.22, the input polynomial is inserted into the coded shift register with $n = k+r$ shift pulses, which present the input of $X^r M(X)$. During this operation, gate G_1 is open and G_2 closed. This allows the shift register to perform the division operation during the required n shift stage. At the same time, the input polynomial is delayed for the period of r shift operations and then the original input information polynomial is shifted out unaltered although delayed. This output process is a combination of r delay plus k message digits, giving a total of n. Hence, at the end of n shifts the remainder is in the register, ready to be shifted out. After n shifts, gate G_1 is closed to prevent any changes to the remainder and gate G_2 is opened so that the remainder can be shifted out with r shift pulses.

The delay of r stages only adds expense to the circuit, and it can be avoided by modifying the encoding circuit, as shown in Fig. 12.23. Observing the step-by-step

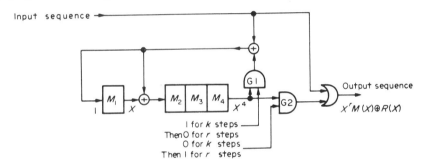

Fig. 12.23. Modified circuit for encoding block codes.

process of the previous circuit, it can be seen that in the first four operations the circuit operates as an ordinary shift register and only after the $r(=4)$ shifts can it start coding. Hence the modification is obtained by automatically multiplying the input by X^r as the input digits are entered into the register. In other words, the information polynomial is inserted near the output which is r stages in front of the previous input. In this modified circuit, after k shift pulses, the register will contain the remainder. Here gate G_1 is opened and G_2 closed for a period of k pulses and then the G_1 is closed and G_2 opened for the following r pulses.

It is not practical to decode the division operation by multiplication, since in one the highest order must be implemented first and in the other the lowest order first. Instead, both the encoder and decoder use the division operation. The decoder is identical with the encoder, and so the same circuit may be used either for coding or decoding, thus minimizing the required hardware. At the receiving end the transmitting polynomial is checked for parity failures. This is implemented by simply entering the received polynomial into the decoder, which consists of both the information message and the check polynomial, viz.

$$T(X) = X^r M(X) \oplus R(X) .$$

It has already been shown that this polynomial is also equal to

$$T(X) = Q(X) \cdot P(X) .$$

Since $Q(X)$ must be unique, there should be no remainder when dividing the received polynomial $T(X)$ by the same generator polynomial $P(X)$. If there should be a remainder by this operation, this will indicate that there are errors. Figure 12.24 shows a decoder circuit which is identical with that shown in Fig. 12.23 except than an extra gate and a flip-flop was added. This extra gate, G_3, is opened after the message and parity polynomial have been received, that is, it is closed for the first n steps and then opened for the following r steps. During these r steps the remainder is pulsed out. If the full remainder is equal to zero, then the flip-flop will remain unchanged. However, if there is any remainder, this output will set the flip-flop indicating an error.

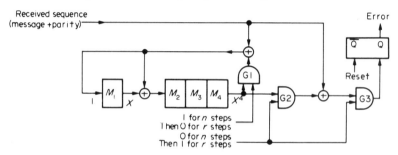

Fig. 12.24. Decoding circuit for block cyclic codes.

Cyclic codes are well suited for error detection, for, by suitably selecting the generator polynomial, they can be designed to detect many combinations of possible errors. Nevertheless, if the transmitted block is divisible by the generator polynomials, there will be no parity check digits. Every cycle code can detect an error burst of the length of r or less. It has already been shown that the minimum distance required to detect an error is $d = 2$. Thus, any generator polynomial with more than one term can detect single errors where the simplest polynomial with more than one term is $X + 1$ (that is, 11 in digit form). Furthermore, any polynomial which is divisible by $X + 1$ will have an even number of coefficients and may detect not only single errors but also any odd number of errors. This can be expressed as $P(X) = (X + 1) \cdot P_1(X)$, where $P_1(X)$ is any other mod-2 polynomial. It can also be shown that among the generator polynomials which are divisible by $X + 1$ are all the forms of polynomials $X^c + 1$. Thus, all the polynomials $X^c + 1$ can also detect an odd number of errors. To detect single and double errors, the length of the code is no greater than exponent c to which $P(X)$ belongs. If there is a double error, the distance between the correct word and the error can be expressed as $X^i + X^j$, where $n > i > j \geqslant 0$. To detect such a double error, $E(X) = X^i + X^j = X^j(1 + X^{j-i})$ must not be divisible by the polynomial generator.

In practical applications various numbers of codes are being used. IBM's Binary Synchronous Communications mode suggests the use of two codes, one of 12 check digits and the other of 16. These polynomials are as follows:

$$P(12) = X^{12} + X^{11} + X^3 + X^2 + X + 1 = (X + 1)(X^{11} + X^2 + 1),$$

$$P(16) = X^{16} + X^{11} + X^2 + 1 = (X + 1)(X^{15} + X + 1).$$

CCITT in their recommendation V41 suggests the use of another type polynomial

$$P(16) = X^{16} + X^{12} + X^5 + 1 = (X + 1)(X^{15} + X^{14} + X^{13} + X^{12} + X^4 + X^3 + X^2 + 1).$$

These two $P(16)$ generator polynomials are shown in Fig. 12.25. They can detect any

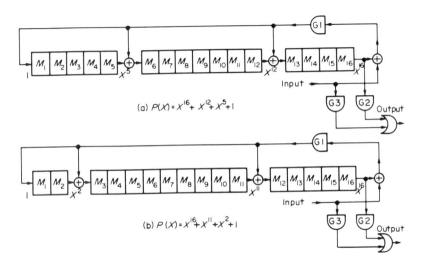

(a) $P(X) = X^{16} + X^{12} + X^5 + 1$

(b) $P(X) = X^{16} + X^{11} + X^2 + 1$

G1 and G3 open for n steps then closed for r steps
G2 closed for n steps then open for r steps

Fig. 12.25. 16-bit cyclic encoders.

error burst not exceeding 16 digits in length. They can also detect all odd numbers of errors and any two errors. These codes give adequate pattern combination of the information blocks. Theoretically, the IBM codes should produce the least number of undetected errors. However, a practical comparison conducted by Standard Telecommunications Laboratories showed the reverse to be the case. Nevertheless, the difference between these two polynomials is negligible.

There are several other important cyclic codes which have not been discussed here. Furthermore, cyclic codes may also be used for error correction applications, such as those put forward by Bose and Chauduri. The further discussion of the cyclic codes in greater detail is, regretfully, beyond the scope of this book.

12.11 REFERENCES

1. Paterson, N.W., *Error Correcting Codes*, the M.I.T. Press, 1961, pp. 1-285.
2. Lin, S., *An Introduction to Error-correcting Codes*, Prentice Hall, Inc., 1970, pp. 1-329.
3. Lucky, R.W., Salz, J. and Weldon, E.J., *Principles of Data Communication*, McGraw Hill, 1968, pp. 277-410.
4. Bennet, W.R. and Davey, J.R., *Data Transmission*, McGraw Hill, 1968, pp. 277-410.
5. Martin, J., *Telecommunication and the Computer*, Prentice Hall, Inc., 1969, pp. 360-390.
6. Martin, J., *Teleprocessing Network Organisation*, Prentice Hall, Inc., 1970, pp. 62-95.
7. Richards, R.K., *Digital Design*, Wiley Interscience, 1971, pp. 178-271.
8. Burton, H.O. and Sullivan, D.D., "Errors and error control", *Proc. IEEE*, Vol. 60, No. 11 (Nov. 1972), pp. 1293-1301.
9. Balkovic, M.D. and Muench, P.E., "Effect of propagation delay caused by satellite circuits, on data communication systems, that use block retransmission for error correction", *ICC Conf. Rec.*, June 1969, pp. 29-36.
10. Kohavi, Z., *Switching and Finite Automatic Theory*, McGraw Hill, 1970, pp. 14-21.
11. Peterson, W.W. and Brown, D.T., "Cycle codes for error detection", *Proc. IEEE*, Vol. 49, No. 1 (Jan. 1961), pp. 228-235.
12. Hagelbarger, D.W., "Recurrent codes: easily mechanized, burst-correcting, binary codes", *The Bell System Tech. Journal* (July 1959), pp. 969-985.
13. Franco, A.G., "Coding for error-free communication", *Electro Technology*, (Jan. 1968), pp. 53-62.
14. Forney, D.G., "Coding and its application in space communications", *IEEE Spectrum* (June 1970), pp. 47-58.
15. Velasco, C.R., "Effect of transmission modes and error-detecting schemes on transmission efficiency", *Telecommunications* (Feb. 1968), pp. 26-36.
16. Wright, E.P.G., "Error correction: relative merits of different methods", *Electrical Communication (ITT)*, Vol. 48, No. 162 (1973), pp. 134-145.
17. Halier, U., Matt, H.J. and Progier, M., "A forward error correction system for heavily disturbed data transmission channels", *The Radio and Electric Engineer*, Vol. 42, No. 12 (Dec. 1972), pp. 523-530.
18. Chien, R.T., "Block-coding techniques for reliable data transmission", *IEEE Trans. Comm. Tech.*, Vol. COM-19, No. 5 (Oct. 1971), pp. 743-751.
19. CCITT, Vol. VIII, Recommendation V.41, 1968.

CHAPTER 13

Secrecy, Security and Privacy

13.1 INTRODUCTION

Secrecy and security are terms usually connoting national and specifically military interests. They have applications, however, far below the national level, ranging from commercial espionage to individual privacy.

It is common practice in most organizations to rely on computers in all matters affecting the personnel. Indeed, it has become almost natural today to have to contact with a computer rather than with a human representative, as the overriding concern in all organizations is increasing efficiency and the maximum saving of time and energy. The computer is undoubtedly a welcome expedient, but we should nevertheless be aware that the computer, an instrument devoid of feeling or sentiment, is insidiously dominating our lives. What is apt to be ignored today is that the increasing exploitation of the storage capacity of the computers may lead to the infringement of individual privacy. Gaining access to personnel files can elicit information about the individual's financial and even his political status.

Another aspect of the question of privacy and the improper use of the individual's file is the possibility that the information in the file is incomplete or even wrong. Administrative decisions regarding the individual could be based on the misleading data within the file, without the individual knowing what is in his personnel file or even that such a file exists. Furthermore, the data contained in the file might only be detected as erroneous by the individual concerned. Indeed, the very existence of such a file may be a potential threat to a man's livelihood.

Despite all the social drawbacks of personnel files, the use of computers and the keeping of individual files are inevitable administrative means. No bank can operate without having on file individual accounts and transactions, and no organization can operate without building up confidential personnel files.

Commercial secrets, when divulged to unauthorized people, could be very damaging to an organization's development, and so they are always confronted with the possible risk of unauthorized persons gaining access to files in order to copy data, change or alter items or even destroy files. In general, confidential information falling into the wrong hands might be exploited for purposes of fraud, embezzlement, treachery, blackmail and above all commercial espionage. The reader may associate espionage with all kinds of sophisticated equipment, but in fact what might be deemed secret information is often left exposed and completely unprotected. In addition, confidential data could be handed out inadvertently.

In the earlier stages of computer installations, the organization kept all the data files in the computer room, and access to the computer was confined to a single location. With the development, however, of time-sharing networks, data can now be inserted into or retrieved from any of the terminals in the system. Despite the fact that all the terminals may belong to the same organization and may be even located in a single building, no organization would surely agree to put all its information files at the disposal of all their employees. Even in the case of those employees allowed access to the files, it is doubtful if they would all be given the right to update the files.

With the growth of sophisticated computer network systems and particularly Command and Control systems, the problem of security and secrecy has thus become more acute. In these network systems, a number of computers share their resources, and so each terminal associated with one of the computers can gain access to files in other computers of the same network. Many organizations may be customers of the same system, and even rival firms may be part of the same network. In the previous chapters, the design aim was directed at the creation of large networks and the sharing of the computer resources among many users, a target which presented both operating and economic benefits. From the security angle, however, the unrestricted sharing of computer resources constitutes a danger. To avert it, the design should therefore make provision for the sharing of some but not all the resources, thus guarding against possible unwarranted access to confidential information.

The sharing of information files in a computer network involves close collaboration between people, and this makes the question of secrecy not only a technical but also a social problem. It is a major design task to produce a system network which is ostensibly for the sharing of files and procedures, but able, nevertheless, to withhold data items from other users of the same system. Any solution to the problem of secrecy is bound to influence in a major way the structure and the mode of operation of the complete system. It must call for the adoption of flexible security measures as an integral part of the original system.

As indicated, the whole subject of security and secrecy is relatively new, and not all the problems involved are known, let alone solved. Indeed, most of the problems are only now coming to light, and in most of the solutions there are still loopholes which favour security leaks. The mere fact that the computer systems are intended to serve a wide range of people is a security risk in itself, since it cannot be expected that all the users will be security minded. The suggested solution should be directed at improving the technical operation of the system rather than enforcing the implementation of security procedures.

The purpose of this chapter is only to point out to the designer the complex problems of security and secrecy and to offer some ideas for consideration.

13.2 SYSTEM THREAT POINTS

It is usual to consider the danger from espionage as the major threat with regards to data information. This is only partly true. There are other dangers which can apply in a computer system, such as malicious destruction of files, or the fraudulent changing of their contents. The most important danger is the inadvertent revealing of data as a result of ignorance, naïvety or over-confidence, for example, where the operator regards the computer data to be in a form that nobody else can read.

The first and most simple solution is to have strict operational procedures. Unfortunately, regulations are not always strictly observed, since the average person is not sufficiently security minded. Most operators scarcely realize the disastrous effect caused by the leakage of private information.

Correct procedures, if kept, may prevent some accidental security leaks, yet no system is assured of being accident proof. Furthermore, procedures alone certainly cannot prevent the unauthorized penetration of the files for the purpose of copying, altering or destroying them. The solution of these problems does not lie in procedures, but in technical hardware and software techniques which can reduce the risk of any such penetration.

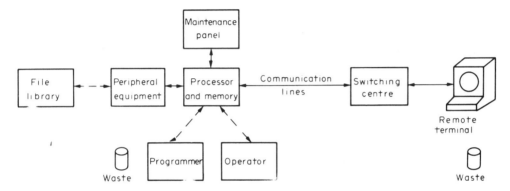

Fig. 13.1. Vital threat points to the security of time-sharing systems.

Before suggesting means of improving the system security, the threat points of the system must be discussed. The threat points in any operating system do not lie only in the terminal which is used as an input–output medium to the system but along the whole system. The vital danger points which may threaten a time-sharing system can be seen in Fig. 13.1, and are evaluated as follows:

(a) *Waste-paper baskets*

All the modern equipment used for espionage, such as bugging devices, cameras, etc., are possibly superfluous, since the waste-paper baskets provide the ideal and easiest means of access to information. Discarded note paper, carbon paper, paper tape, copy ink ribbons, punched cards, etc., are an excellent source of information, and so are computer forms and data cards. To obviate the danger two security precautions should be taken. The first is to have strict procedures whereby all the waste is destroyed. The second precaution is, where possible, to keep most of the data in code, or, at least, to separate the data from its heading. The data can be printed without the real account number, which can be added at a later stage.

(b) *Terminals*

There are two types of terminals, those that provide hard copies and those that provide only soft copies. In the first category are the typewriters and the printers, while in the second category are the various displays. It is obvious that most of the security risks are with the first category; nevertheless, both categories present a serious security threat to the system. The terminals are obviously the most vulnerable access points in the operating system. To prevent access to the data files at the terminals by unauthorized people recourse must be had to special identification codes for use as entry keys both by individuals and by means of the terminal itself. The system may then identify the user by the special entry key he is provided with. Another source of security threat applicable to both types of terminals is negligence through which opportunities are created for the casual observer to glance at and extract the sensitive information. Another point that should be considered is that hidden recorders may be attached to the terminal and all the transactions recorded without the user being aware of it.

(c) *Switching centres*

Selection of the various terminals may be by circuit or by message switching which may be an integral part either of the system or of the public telephone network. Both categories may present security threats to the system. The switching centre may accidentally connect the computer to the wrong terminal, and thus it may print out a classified file to an unauthorized terminal. Another security threat is the cross-talk which goes on in every switching centre. Cross-talk can be reduced but it is difficult to assure its elimination. The latter threat is much more acute when the data is transferred through the public telephone network. The switching centre could also present another security risk, since it can present an ideal spot for the attachment of taps to the switching unit of the terminal. Preventing all these threats can only be implemented by cyphering the data transmitted through the centre. No secret data should be passed through the switching channels unless the data is obscured or cyphered. Code entry words may prevent active unauthorized entry but will not protect against eavesdropping. The only effective counter measure is for all the data to be in cypher.

(d) *Communication links*

Communication links present the same security threats as switching centres, that is, they are both susceptible to cross-talk and wire tapping. These leak risks are most difficult to prevent, since the communication lines are generally long and open. The only means of circumventing the risks is cyphering all the data flowing in the lines. Each terminal needs to be supplied with a unique cypher decoder, so that only the appropriate terminal will be able to decypher the information.

(e) *Processor memory*

As it is extremely difficult to process cyphered data, the processing must deal only with clear data. The use of cyphers for processing must perforce be restricted to such items as file indicators and names.

Each processor should be provided with two check monitors, one for the operation of the terminal and the other for the use of operators. The first monitor, containing the code tables, must be protected to allow only restricted access. The second monitor may be unprotected, to enable the operator to supervise the system operation. Admittedly, where there is an unprotected monitor in the system, there is always the danger that the operator may accidentally or deliberately switch the monitors.

The major security risk in the processor is that the monitors and the code translation tables are revealed. This risk can be generally reduced by keeping the tables under 'execute' only, and furnishing the processor with hardware and software measures to prevent unauthorized entry. The hardware protective measures must be an integral part of the system, while the special software control techniques are for control over access. All the protective means noted here must be designed into the system and not left for later addition.

(f) *Operator*

The operator has the greatest implementation power in the system, since he is generally provided with means of access to the data. Both he and the system programmer, as well as the maintenance personnel, have special privileges in any operating system which they may misuse. They thus constitute a potential threat to the system security. By appropriate instructions they may command the computer to print out data from files which they are not authorized to handle. Means must therefore be introduced to neutralize this risk, although the intelligent operator may still find means of by-passing any precautions. The greatest fear of all is that the operator may reveal the protective measures used in the system and in particular the translator tables. The system must be so organized that the operator, under normal conditions, has no access to the code translation tables. Another danger to the system lies in the possibility that the operator's errors or neglect may cause the accidental destruction of the files.

(g) *Maintenance panel*

The maintenance personnel generally have restricted access to the user's data files; nevertheless, they are provided with the tools for gaining access. While attempting to diagnose a fault they might easily discover the protective control precautions of the system, and if they wished, rewire the processor hardware to make it appear secure when in fact the security precautions have been by-passed. As the maintenance personnel must have a thorough knowledge of all the circuit hardware, including all the system's security precautions, they can perform unauthorized modifications without their being detected by the other system operators. In consequence, the maintenance personnel are potentially a greater threat to the system operation than the operators. It is in their power to introduce new programmes (which they would call diagnostic programmes) and while so doing explore the protective system and/or gain access to the user's files. The only way to prevent the maintenance personnel from disabling the protective devices, or at least to make this more difficult, is to have the protective hardware devices as an integral part of the operating system.

(h) *System peripherals*

The processor generally handles uncyphered data, and so do the centre peripherals. This being the case, all the data being used by peripherals could also present a threat to the security of the system. The only precaution that is suggested here is to keep the files with coded headings and not to use their actual names. Furthermore, operation procedures must be strictly enforced to prevent any unauthorized access to the peripherals.

The line printer is usually a main peripheral equipment near the processor. Hard copies are the easiest means of obtaining direct knowledge of private or confidential data. Salaries and accounts transactions are all printed out, thus providing all the vital private details. This gives the operators the power of knowing secrets concerning individuals. The danger may be reduced by using special double sheeted paper, one page being blank while the other has the print. As the paper leaves the line printer, both papers are stuck together with the upper sheet obscuring the data and with only the name of the account appearing on the outer sheet. The data can be revealed only by peeling off the outer cover along the perforated borders, but then the sheets cannot be replaced again.

(i) *File libraries*

File libraries in use for long-time storage call for special protective measures. Apart from possible loss or misplacement of files, there is the danger of penetration for illicit purposes. To guard against the latter contingency, the libraries should be kept under lock and key or, preferably, under the constant surveillance of appointed personnel. As a further precautionary measure, all removable storage material should be kept in cypher. Then even if they are stolen or copied, the cypher would have first to be broken. The libraries need to be guarded not only from outsiders but also from the organization employees, who though familiar with the system's programmes are still restricted in their access to the users' programmes.

(j) *System programmers*

The system programmer is the person most familiar with all the protective measures in the system. He is responsible for the system programme design and implementation, as well as for maintaining the system-control programmes. Although the programmer may not be directly associated with the secret material, he still has the know-how for reaching it. It is in his power to introduce private inputs to the system which can by-pass the software protective measures. Moreover, in cooperation with the system engineer he can employ highly sophisticated means for penetrating the system, using techniques which may allay suspicion and enable him to act without being detected. Another possible though remote threat is the revealing of the protective measures by a system programmer eager to publicize his ingenuity in a technical journal.

The threat points listed above could be arranged in the following categories: (a) remote access; (b) internal system access and (c) communication access. The threats themselves would also be grouped in three categories, viz. (a) unauthorized entries; (b) operators' and programme errors and (c) hardware failures. The main threat to the security of the system is from illicit entries, which could be made for the following purposes: destruction of a complete file or even the complete system, copying a file or parts of it, theft, deletion of items, modifications to items in the file or to the complete file, insertion of false data, or simply unauthorized examination of the files. Operators' errors could be caused by accident, negligence or simply by ignorance. Hardware failures could only be prevented by a reliable design, as depicted in the following chapter. When a failure does occur it must be fail-safe in respect to security.

13.3 ENTRANCE KEYS AND PASSWORDS

One of the most common solutions to secrecy and security problems is the use of special entry codes as means of access to the data files. By the use of these special codes, which may be checked either by hardware and/or software techniques, access to the files may be limited to the legitimate entry of authorized personnel only. However, the entry problem is far more complicated than the simple granting of permission, since not all those that have access to a file may regard it as being at their disposal. That is to say, not all legitimate access to a file implies a right to retrieve all the items in the file. Furthermore, permission of entry for the purpose of glancing through the data in the file does not grant the additional privilege of changing the data items. All these points make the solution by the use of entry codes much more complicated than the issue as presented here and raises the question whether entry codes are indeed sufficient.

The entry code is in effect the authorization for gaining access to a particular file. It is the means whereby the computer may identify the user and allow him to enter the system. The granting of an entry is not enough, since the system must also recognize and decide what operations the user is allowed to perform after the entry has been granted. Some of the problems which must be checked before final permission is granted are listed below.

  (a) Recognizing the authorized user by means of a unique identification code.
  (b) To what files can the user have legitimate access?
  (c) What data can the user retrieve? Are any of the items in the file still restricted to particular users?
  (d) By whom and how can the data be updated? If all the callers can update the files, then the data in the file becomes unreliable. New data may come from a number of sources, but only one source should be allowed to update the files after checking the validity of the input information.
  (e) Who is allowed to change the user's operating procedure, and how can it be implemented without infringing the security precautions? The implementation must only be performed from a single point, preferably from the centre, using both hardware and software techniques. No user or terminal should be granted permission to change procedures or introduce major operating programmes unless special precautions have been introduced to ensure that it will not affect other users.
  (f) How can the operating procedures be changed, and who is responsible for the programmes which are common to many users?
  (g) How can programmes be shared by a number of users and at the same time be restricted for the use of others?

The above points may be divided into two categories: (a) limiting the access to the users' programmes, which is a secrecy problem, and (b) limiting access to system programmes, which is a security problem. Limiting access to users' programmes could be achieved by the use of special entry codes. However, no access should be allowed to system programmes from any remote terminal. That is, the code word should have limited validity, granting ability to reach a particular item but not the power to change procedures or alter any system programmes. The latter may be implemented only by direct access at the centre with the aid of the supervisor. No users may be allowed by the system to violate the established procedures or be granted the tools for entry into the full system.

Special unique codes are required for each terminal, for each user and possibly for each entry operation, and different type codes are to be used for read and write access operations. These codes may be in the form of passwords. The system will deny access to those who do not possess the appropriate password, to those who misuse it or to those who take advantage of their passwords to try to enter restricted areas.

Each file should be headed with a code identifying the list of those allowed access to it and the parts in the file which must be further restricted. Special tables are constructed in the computer to record the passwords and their authorization, together with access file codes and the counter references. These tables register the full list of authorized users and indicate what files they can request and have access to and what files they can legitimately update or modify. Granting access can then only be made by cross reference of this list with the individual code of the file. In this way, tampering with the table list of user identification codes would still not grant entry to the file if the two lists do not coincide. The actual tables are supplied by a combination of hardware and software to prevent illicit changes by programmers. Furthermore, the tables must be specially protected from any accidental or deliberate access, as the whole system security would become useless if the tables used were made public. Wrong access to these tables can be made even more difficult by imbedding the table operating programmes under 'execute-only' type of memory protection.

There are many code password types: (a) single occasion passwords, supplied as required to all the users of a particular system; (b) unique passwords, restricted to a single user and (c) functional passwords which indicate the user's security classification. All these codes become invalid when they are lost or stolen. Another and preferable group of code password types are (d) changeable and (e) random code passwords. In these types, a different code combination is adopted for each access operation. Here every user may be provided with a set of permissible codes, each presenting a sequence of random numbers. After the use of each code, it is cancelled and the one following is selected. One-time random passwords are useful only if the transactions are few, as the procedure can become irksome if the users require many short entries. Other types of changeable passwords include the use of dates, a running sequence of numbers or a special code which may be changed daily. A disadvantage of all passwords is that the system is over-sensitive to false entries and may not discriminate between human errors and unauthorized entries, and will thus be liable to record a false entry as a security infringement.

The use of fixed passwords which are exclusive to each user can be inadvertently disclosed by the user, thus making access available to unauthorized persons. Many people write the code password on a slip of paper which can easily be lost. Others leave it jotted down near the terminal, where it is easily available to anyone who has access to the terminal. Furthermore, as a mnemonic device, code passwords are usually selected to relate to the user's name or to the file subject and this also facilitates the breaking of the codes.

With the use of the code password a special procedure must be observed whereby the user first signs in by stating his unique password. Repetitive transmission of the password with each block transaction or request unduly complicates the operation, and so some systems may only require an abbreviated code with each block. Alternatively, access may be permitted on an indefinite basis, which is granted after the user has "signed-in", but this will require a "signed-out" code which is implemented when the transaction is completed. Here, however, there is a danger that the user may leave the terminal without signing off, thus making the terminal available to unauthorized personnel.

Some systems use the terminal keyboard itself as a means of automatically sending in a unique password which is associated with a particular terminal. This may be extended to include a skeleton key which is inserted into the terminal. These techniques, however, are not advisable for use by themselves, since the process may complicate the changing of terminals for maintenance purposes, and in any case it does not offer secure precaution.

Another entrance means could be a specially coded key in the form of a magnetic strip or punched card. Each user is provided with this coded card which contains a sequence of binary codes unique to each user. The coded key is inserted into a special slot which causes the transmission of the user's specific code followed by the terminal identification. A special advantage of the coded key is that the sign-off procedure is simplified. The snag here is that the keys may be lost or stolen, with probable delays till the user realizes and reports their loss. To reduce the danger of unauthorized use of the keys, the validity of the keys may be restricted to particular terminals where both user and terminal codes are checked and cross-checked.

All these precautions still leave much room for breaches of security. An improved scheme on similar lines is the extended handshake. In this scheme both coded keys and passwords are used, sometimes associated with a particular terminal. The user inserts his magnetic coded key into the particular terminal which can verify whether the key (not the user) and the terminal have a legitimate right of access. Once the system has verified his identification, the user is notified and can proceed by transmitting his unique password. This handshake could be extended to include a series of checks which may include current passwords and a random selected code. After each entry, the system must verify it by matching it to a given table, and only then can the user proceed. Once the caller is fully identified, he must ask for clearance to grant him access to a specific file or job. This too must be verified before he may proceed with his specific project.

Entry keys using physical characteristics such as voice or fingertips have as yet not proved satisfactory. The system may reject a legitimate user's request when he may merely be suffering from a common cold which has distorted his vocal chords. Finger-prints too are not yet absolutely reliable, as they may be reproduced on a rubber band and in this form successfully by-pass the system security precautions.

13.4 FILE SECURITY AND SECRECY

The main design task of security and secrecy in computer systems must be directed at guarding the files, since they contain all the relevant secret data which could be damaging when divulged. The problem of keeping the files secure is a double one: guarding the files when they are in a remote storage area (i.e. off line) and preventing access when the files are in the active system (i.e. on line).

When limiting access to the files, it is not enough to define the list of authorized people who can have access to a particular file; it is necessary also to limit the information items that can be extracted by each authorized person. Each file may contain information of different levels of sensitivity or classification. Therefore, means are required to control access not only to the files but also to each level.

The off-line files are generally kept in special libraries in the form of paper or magnetic tape or discs. All these storage media of the files are removable, and as such are more vulnerable to loss or theft. The natural precaution is to keep the library under lock and key, but this, although essential, is not adequate. Labels could be lost or accidentally switched. It is quite a common mistake for the library to hand out the wrong file. This could be disastrous if the information in the file were inadvertently handed over to a competitive organization.

All remote files must have a software coded heading which will identify not only the file but also its legitimate users. Thus, if the file is accidentally or intentionally switched, the computer could still recognize this fact and prevent its wrongful use. The computer will hold a protection table listing all those who can gain access to a particular file. This table is separate from the list of authorized people who can gain access into the system. Furthermore, the file access list must not be in the file heading but in the computer hardware. The code at the heading of the file is checked by the table and cross-checked with the user's request. Only if they tally can the file be processed to the users' request.

The coding heading may protect the file when it is on-line within the active system. However, this is not enough for off-line storage. The files are liable to be lost, stolen or simply discarded after use, and anyone who gets his hands on these could then run them on another computer by by-passing the heading.

For the secrecy of all long-term files it is not enough for the file to have a coded heading: all the data must be cyphered, so that, if the files are lost or stolen, the data on them would be useless unless the cyphering key were also available. The coder and decoder of the cyphered information should be unique for each user and be held within the active computing centre (cyphering techniques will be discussed in Section 13.6). Any data leaving or entering the computer, either by means of the communication links or for remote storage, should be preferably in cyphered form. This will also prevent any illicit infiltration to change items on the files for the purpose of fraud or embezzlement. Without knowing the cypher, no changes can be made in the data (for example by punching extra holes on the cards) without their being recognized. Furthermore, besides the cyphering of the data, error-detection codes should be added so that any changes in the data format are also detected.

The security of the long-term files is a different problem from the one discussed above. Means must be introduced to ensure that the absence of a file will not adversely affect the operation of the system. In spite of all precautions it is almost inevitable that accidents will occur from time to time. As far as possible this contingency must be anticipated and means adopted to neutralize their damaging effects. The safest insurance in this connection is keeping the files in duplicate or triplicate and distributing them between different storage centres and locations. Furthermore, thought should be directed to finding means of reconstructing the files in on-line real-time systems once they have been damaged or destroyed.

The secrecy and security of the files while they are in the centre processor is a much more difficult problem than when they are in a storage location, for in the former case although the files cannot be stolen or lost they are vulnerable to unauthorized entry and could be then copied or damaged. Restricted files must be guarded against all unauthorized personnel, including the operators. This is not a simple matter, since the data cannot be processed while being cyphered. Using code passwords is one of the means of precaution but it is far from perfect. Another means of guarding file entry is the use of code passwords, but this, too, has its limitations, especially against the system operators. A third means of guarding against such infiltration is to keep all headings in code so that the data cannot be referred to a particular person or job. The translation of the code could be handled by the user himself instead of by the centre. This solution is far from perfect but it does present a reasonable measure of security.

Limiting access to a particular file may be a good procedure in the case where a restricted list of authorized people can be drawn up. However, in sophisticated resource-sharing systems and also in Command and Control system, access is a more complex matter, since different competitive organizations may have to operate together, sharing some of the files but with other files restricted. In addition, the various subsystems must be able to converse intelligently about their data contents without any codes or cyphers disturbing this communication. In other words, the system must be able to allow those items intended for transfer to be revealed while at the same time keeping the same items secret from other users of the system. This may require that the same file will have to be cyphered by different techniques so that it could be transferred to different users. It may also require that the same information item will have to come under different code headings so that it can be extracted from the various subsystems without revealing the rest of the file.

13.5 HARDWARE AND SOFTWARE PRECAUTIONS

All the precautions discussed in the previous sections still leave loopholes for unauthorized entries by operators and maintenance personnel. These loopholes must be effectively plugged. Although their implementation will be effected by software techniques, all the precautions have to be introduced into the equipment, the features of which must therefore be considered as an integral part of the system.

The computer must have adequate memory protection, partitioned requests and privileged instructions. The set of privileged instructions supervise the system operation and only they can change any memory-protection barriers. The storage limits can be loaded by the Executive system, which can thus control the operation of a programme currently in execution. This protection is effected by the ability of the Executive to stop a programme at a point of error or violation.

An important precaution is the separating of duties associated with the system. This not only prevents security leaks but also reduces the risk of operating errors. In this technique each operator may also have his own password, which may allow access only to the operating programmes but not to any users' programme. Access to the users' programme may only be allowed either from the relevant terminal or from the centre by someone who possesses the correct password.

The most important precaution of all is the keeping of a complete record of the systems' operations. That is to say, the full record of the machine performance must be continuously logged, thus recording all the systems' transactions. Each entry is recorded separately and can later be cross-checked to reveal if there has been an unauthorized entry, in addition to maintenance and other system transactions. Any attempts of the operator or the maintenance staff to by-pass the protective control would be recorded and their violations could thus be detected. Furthermore, each transaction is measured by the operation time length, thus reducing the likelihood of a protection by-pass for long periods of time by attempts on the part of the operators or maintenance crews.

Keeping a record of events logs all the computer transactions and all the input/output operations. With the aid of this record it would be possible to ascertain which terminal performed an operation and to what extent there was an attempt to perform an unauthorized operation. The same record could also be used to collect data about all the file transactions for statistical purposes and future security classification.

Any violation of the security precautions must cause an alarm to be sounded in the system. In response, the first action that should be taken is to lock the terminal so that no damage can be caused to the system. Then, an alarm indication must be sounded in the centre for the operator to intervene. The operator will check the cause of the unauthorized entry and only then release the terminal for further operation. At the terminal the user must start all over again by inserting his password. This, of course, is only done after the operator has ascertained that there was an error rather than an illegal operation. It is essential that the operator should not have the ability to provide a user or terminal with means to by-pass the individual password entry code.

The monitor recording transactions must log all security violations, that is, not only unauthorized entry trials but also unauthorized transactions. This restriction is essential to prevent fraud where an authorized person tries to gain access to a confidential file or to change items in a file which he is allowed only to retrieve. This violation, taking advantage of a legitimate access to the system so as to insert an unauthorized request, must be stopped by sounding the major alarms the moment it is detected.

The system can easily eliminate unauthorized entry to the system but it is difficult to restrict the legitimate user from entering the unauthorized areas. Once a user gains access to the system he may make use of sophisticated programmes in order to browse through prohibited files. A precaution against this type of infiltration is furnished by both hardware and software techniques which can impose processing restrictions in addition to the memory-protection features. Furthermore, special programmes must be introduced to recognize suspicious requests, even those which may eventually prove to be legitimate. Such requests, for example, may be for a print out of the full file or for the erasure of the complete file. Where there is such a request, other procedures should be followed. As the request may be legitimate there is no point in locking the terminal, although as a cautionary measure it may well be to request the intervention of the operator before the job-process request is granted.

Special programmes must also be used to check the validity of the security precautions, in order to ensure that they are still in operation and have not been by-passed. These programmes should perform periodical checks to see if there have been any hardware or software violations introduced into the system. These checks should not be limited to periodic inspection but should be performed also after each maintenance repair or modification.

13.6 CYPHERING TECHNIQUES

It has been suggested in this chapter that all secret and/or confidential data flowing in the communication channels should be cyphered. It was further suggested that all data files placed on removable storage media should also be in cypher. The cyphering of the data is intended to prevent the extraction of any useful data from the files should they fall into unauthorized hands.

Before a particular cyphering technique is selected it is essential to define the type of penetration it is intended to prevent. There are simple cyphering techniques which may protect against the casual browse through the data, while there are highly complicated techniques which may require a computer analysis operation to break the code. For top-secret data completely unbreakable codes are essential. Such unbreakable cyphering techniques are based on random codes which are used only once. One-time codes can, of course, be extremely burdensome for commercial organizations which transmit only confidential data.

It is not possible to cover the entire subject of cyphering techniques; but attention must be drawn to certain techniques of cyphers which are based on modulo-2 arithmetic. One of the first cyphering methods is the one called the VERNAM technique, named after its inventor who designed it in 1917. In this technique the information word is linked with a cypher code word. That is, the two words are added according to modulo-2 addition principles. This operation can be appraised from the demonstration of the following example:

Information word	1 1 0 1 0
Cypher code	0 1 1 0 1
Coded information	1 0 1 1 1

The original information may be reobtained again by using the same key for decyphering, that is, by mod-2 adding of the coded information and the cypher code, as seen from the following example:

Coded information	1 0 1 1 1
Cypher code	0 1 1 0 1
Original information	1 1 0 1 0

This technique, as described, could easily be broken, but it becomes more difficult if a set of different cyphered codes is used. Vernam created a semi-random code by using an eight feet length of closed loop punched paper tape. Each character in the tape presented a different cyphering code. J.O. Mouborgne, of the U.S. Army, extended the Vernam technique, using one-time code tapes, where the cypher code was derived from a random noise source. This technique is shown in Fig. 13.2, and it can be gathered from it that the system is reversible. The major disadvantage of one-time tapes is the enormous volume of

Fig. 13.2. One time tape cypher—based on random coding.

tapes required. Both the Vernam and Mouborgne techniques present operational difficulties in synchronizing both ends of the system. For completely successful operations, the right cyphering code word has to be applied simultaneously both at the receiver and transmitter, and this becomes a major difficulty when one-time tapes are used.

A pseudo-random system could be based on two separate cyphering codes, that is, the information is added twice, as shown in Fig. 13.3. This technique is more similar to the

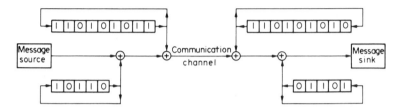

Fig. 13.3. Pseudorandom cypher—based on cyclic registers.

Vernam than to the Mauborgne technique, since the number of cyphered codes is limited. However, instead of using tape, long cyclic shift registers may be used, where their combination presents a semi-random number of codes. To increase the combination response of the system, the two registers are made of unequal and disjoint length. The longer the length of the shift registers, the better the cypher. Furthermore, the contents of the data sequence buffered in the register could be changed at frequent intervals. Although long registers can represent a magnitude of random combinations, these combinations must repeat themselves after a number of runs. Where the traffic volume is large, an analysis of the coded information could reveal the cyphering key. It is possible to increase the capabilities of this technique by automatically changing the contents of the registers when all the combinations of the two registers have been used up. If the length of both registers is n and m respectively, then the number of operations is equal to $n \times m$ provided $n \neq m$. Since the sequence repeats itself and its combinations can be predictable, the technique is known as a pseudorandom sequence.

The double modulo 2 additions could be extended to include a number of addition operations. In fact, a single cyclic register could be used where the modulo 2 addition operations are in the feedback circuit. These cyclic registers were studied in the previous chapter for the use of error control. It was shown that a shift register with n stages may produce $2^n - 1$ different combinations, provided a maximum length of register is used. Although this technique produces a fixed number of combinations, and thus also produces a pseudorandom sequence, the number of combinations obtained for the same number of register stages is far greater. A shift register of 20 stages can produce a sequence of 1,048,575 different combinations, and this number may be doubled with each extra stage added, where for 30 stages, 1,073,741,823 combinations are possible.

The cyphering operation is performed by the mod-2 multiplication of the input polynomial by the generator polynomial, as shown in Fig. 13.4. This type of circuit was suggested in the previous chapter for coding convolutional data. The decyphering operation can be simply performed by dividing the cyphered polynomial by the same generator polynomial. However, this operation can create certain difficulties, because multiplication must start from the lowest order and division from the highest. For this reason, the cyphered information must be buffered and then reversed. Alternatively, the generator polynomial can be reversed for the multiplication operation and thus give the identical result. Table 13.1 shows the step-by-step progress of the circuit shown in Fig. 13.4 both for the cypher and the decypher.

TABLE 13.1

Input data	Multiplication				Cyphered data	Division				Clear data
	M_1	M_2	M_3	M_4		M_1	M_2	M_3	M_4	
1	0	0	0	0	0	0	0	0	0	
0	1	1	0	0	1	0	0	0	0	
1	0	1	1	0	0	1	0	0	0	
0	1	1	1	1	1	0	1	0	0	
1	0	1	1	1	1	1	0	1	0	
0	1	1	1	1	0	1	1	0	1	1
1	0	1	1	1	1	1	0	1	0	0
0	1	1	1	1	0	1	1	0	1	1
1	0	1	1	1	1	1	0	1	0	0
1	1	1	1	1	0	1	1	0	1	1
	1	0	1	1	0	1	0	1	0	0
	0	1	0	1	1	0	1	0	1	1
	0	0	1	0	1	0	1	1	0	0
	0	0	0	1	0	1	0	1	1	1
	0	0	0	0	1	1	0	0	1	1
	0	0	0	0	0	0	0	0	0	

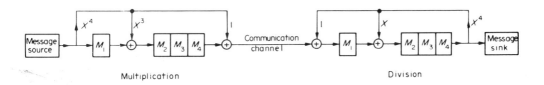

Fig. 13.4. Pseudorandom cypher—based on cyclic codes.

From Table 13.1 it can be seen that this technique could be used to cypher continuous streams of data as with convolutional data. There is a delay between the input and output which is inevitable with a register circuit.

The reader might wonder whether the division operation could be used as a cypher. This is not feasible, because in the event of a remainder it would have to be transmitted to permit proper decyphering.

The generator polynomial could be changed at frequent intervals to present different feedback combinations and thus present a different sequence of codes. A generalized shift register could be used whereby different combinations are selected, as shown in Fig. 13.5. The feedback combinations could be selected by the computer with the aid of an address register. The addresses are so arranged that they will select only maximum length codes. The same circuit is used both for multiplication and division and is selected by suitable

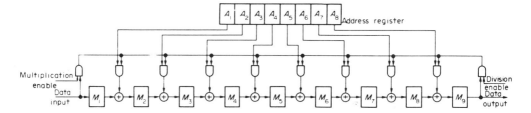

Fig. 13.5. Generalized cypher machine based on cyclic codes.

'enable' gates. For uniformity of the two circuits an extra delay stage is added, that is, the first stage for division and the last stage for multiplication. This extra stage will not, however, affect the operation of the cypher.

Despite the fact that these cyphering techniques are similar to error-control principles, the two do not possess any common properties. In the first place, if any transmission errors creep into the cyphered data, it may make the data undecipherable. Secondly, in all error-control techniques, the encoded digits were followed by clear uncoded digits, while in cypher techniques only the coded digits are transmitted. This problem, of errors creeping into the cyphered sequence, is common to all types of cyphering techniques. For this reason it is advisable for all communication techniques to introduce some means of error control besides cyphering the information. Any error control used must be applied separately and introduced into the communication channel only after the information has been cyphered.

13.7 CONCLUDING COMMENTS

It is obvious from this chapter that all systems must introduce some means of precautions to ensure the security and privacy of the system. Security may add 5-10% to the overall cost of the system but, however relatively high, it is an essential item in all systems. In some systems, the security is needed to guard against embezzlement and fraud, while in others it is introduced to guard against commercial espionage. Even in systems which may have no fear of privacy violations, security of a sort is required to guard against accidental distortions of the file.

The major question to be resolved is how far the system should go in introducing the security and secrecy precautions? One point that must always be borne in mind: the system must be truly efficient, with the precautions introduced not unduly affecting the working arrangement, and in particular, the real-time performance of the system.

In general, security precautions should be simple and easy to implement, and still achieve maximum efficiency in preventing unauthorized violation. These precautions must be within the system and not added later as an extra safeguard. This means in effect that the precautions must be based on the systems' hardware, although it may be operated by software instructions. Furthermore, no precautions should cause undue delays in implementing the user's requests. The checking of the violation possibility should be performed expeditiously, if possible without the user being aware of any checks.

When violations are detected, the terminal must be locked and the operator informed of the security breach. However, the system must not be so sensitive as to cause the locking of the terminal when a legitimate user has performed a minor error in using his password. In such a case a second trial should be allowed before any action is taken. Nevertheless, this error must be recorded in the log, to be double checked later.

The levels of precautions introduced into the system should accord with the danger of possible penetration. Not all the threat points in the system require the utmost precautions. Where the system holds data the disclosure of which could cause serious damage, major precautions should be enforced. If these precautions create delays, alternative means should be looked for, and the delays should be checked to ascertain whether they were unavoidable. Nevertheless, if no solution is found, the precaution must be enforced, despite the delays, since the disclosure of the information would cost the organization far more than the cost of the operation disturbances. On the other hand, where there is no danger of massive penetration, a relatively low-quality protection technique may be applied.

All systems must introduce security precautions against operational accidents and negligence. The body responsible for the systems precautions must also be responsible for all the safeguards introduced into the system and must ensure that no loopholes are left whereby employees can penetrate the system and gain access to their customers' files. Supervisory procedures must be employed in all systems which check and record all violation attempts by the personnel at the system centre. These precautions must be distinct from the guard against unauthorized access from any users' terminal.

When designing secrecy and security precautions, the designer should first prepare a list of all the specific threats to the system. Then he should define the type of information data the disclosure of which could cause damage to the organization. He should evaluate all the alternative precaution techniques so as to assess the cost of implementing them relative to the apparent threats and to the danger of information disclosure. In other words, as in all system designs, security and secrecy must be studied with regards to the cost effectiveness of the solution to be used. There is no point in introducing an expensive precaution if the trade-off analysis shows that the disclosure is of no major consequence.

Cost effectiveness can only be taken up to a certain point. The disclosure of individual private files might be of no consequence in the operations of the organization. It might rather be considered that the expense of the precautions involved does not justify the outlay. Nevertheless, these precautions must be introduced. Indeed, there ought to be an international law enforcing the provision that these precautions should be embedded in all systems dealing with private information.

13.8 REFERENCES

1. Ware, W.H., "Security and privacy in computer systems", *AFIPS, Spring Joint Computer Conference*, Vol. 30 (1967), pp. 279-282.
2. Ware, W.H., "Security and privacy: similarities and differences", *AFIPS, Spring Joint Computer Conference*, Vol. 30 (1967), pp. 287-290.
3. Peters, B., "Security considerations in a multi-programmed computer system", *AFIPS, Spring Joint Computer Conference*, Vol. 30 (1967), pp. 283-286.
4. Petersen, H.E. and Turn, R., "System implications of information privacy", *AFIPS, Spring Joint Computer Conference*, Vol. 30 (1967), pp. 291-300.
5. Skatrud, R.O., "A consideration of the application of cryptographic techniques to data processing", *AFIPS, Fall Joint Computer Conference*, Vol. 35 (1969), pp. 111-117.
6. Weissman, C., "Security controls in the ADEPT-50 time sharing system", *AFIPS, Fall Joint Computer Conference*, Vol. 35 (1969), pp. 119-133.

7. Comber, E.V., "Management of confidential information", *AFIPS, Fall Joint Computer Conference*, Vol. 35 (1969), pp. 135-143.
8. Colimeyer, A.J., "Data base management in a multi-access environment", *Computer*, Vol. 4, No. 6 (Nov./Dec. 1971), pp. 36-46.
9. Beardsley, C.W., "Is your computer insecure?", *IEEE Spectrum* (Jan. 1972), pp. 67-78.
10. Feistel, H., "Cryptography and computer privacy", *Scientific American*, Vol. 228, No. 5 (May 1973), pp. 15-23.
11. Fand, R.M., "On the social role of computer communications", *Proc. IEEE*, Vol. 60, No. 11 (Nov. 1972), pp. 1249-1253.
12. Donn, E.S., "Secure your digital data", *The Electronic Engineer* (May 1972), pp. 5-7.
13. Savage, J.E., "Some simple self-synchronizing digital data scramblers", *The Bell System Technical Journal* (Feb. 1967), pp. 449-487.
14. Peck, P.L., "Achieving security and privacy of information in an on line data processing environment", *ON-Line 72 Conference Proceeding, Brunel University, 4-7 September 1972*, Vol. 2, pp. 107-129.
15. Enslow, P.H., "Non-technical issues in network designs – economic, legal, social and other considerations", *Computer*, Vol. 6, No. 8 (Aug. 1973), pp. 21-30.

CHAPTER 14

Reliability and Maintainability

14.1 INTRODUCTION

With most electronic equipment the designer is concerned with equipment 'reliability', by which is meant the probability that the equipment will perform satisfactorily for a given period of time under specified conditions. With Command and Control systems, the necessity of reliability is far more demanding. The system reliability performance can no longer be defined for a given period of time, because even a momentary failure could be serious. Most Command and Control systems operate with on-line connection to the sensors, which must respond in real-time; for this reason it is not the system reliability which is the most important consideration but the 'maintainability' of the system, by which is meant the ability of the system to maintain its service operations no matter what faults occur.

The reliability design in Command and Control systems calls for the use of highly reliable components, well-proven circuits and redundant methods. However, owing to their extremely complex nature, Command and Control operations cannot be guaranteed against failures by reliance only on the high reliability of their individual components and modules. For maintainability design, the whole system structure has to be considered to ensure non-occurrence of operational errors. Redundancy techniques used for achieving high reliability are not sufficient, and additional techniques must be introduced, such as automatic failure detection and automatic transfer of units. The maintainability objective calls for the continuity of the system serviceability even when the system is not completely free of errors. In other words, the system must remain fully operative even when some of its units are at a standstill due to failures. What is essential here is to prevent further unit failures by chain reaction, and for this to be managed the fault must be located at once and then isolated for a quick repair response.

The system must continue its operational functions not only in the event of a single malfunction but even of a number of software errors and hardware failures. A failure in one of the units must be detected before it is given a chance to propagate and cause further errors involving the other units. To achieve maintainability, there must be sufficient checking facilities built into the hardware of the system.

The capability of a system to perform its prescribed functions satisfactorily is often referred to as the 'availability' of the system. The characteristics of a Command and Control system, which always requires very high availability, is determined by the time allowed for the fault detection, fault location and the automatic change over of the units.

The process of applying tests to determine whether the system is fault free is generally known as fault-detection, and that of identifying the failure fixation within the unit is known as fault location. The adoption of both processes is referred to as fault-diagnosis.

We use the term 'fault-tolerant' for the ability of the system to continue its operational operations in the presence of a failure. While fault-diagnosis is achieved by special techniques generally introduced at the circuit level, fault-tolerant operation is reached by means of protective redundancy introduced at the module level.

Much study has been directed in the last few years to fault-diagnosis and fault-tolerant computing techniques, but although many articles have been published on the maintainability of airborne systems, very little has been published on the average Command and Control systems for commercial use. Nevertheless, no real-time on-line system configuration should be designed without introducing redundancy and maintainability techniques. The purpose of this chapter is to draw attention to the importance of the subject and to note its contribution to the system's prolonged performance. It is not intended to serve as a guide to designing fault-diagnosis techniques within a system, since system designers are not required to go below the module level to the circuit and component level.

No Command and Control system should be designed without including fault tolerant techniques, and this is possible only if fault-diagnosis techniques are adopted. The Command and Control customers must get the full system services under all conditions, without the probability of failures interfering with their operation. The fault-tolerant technique is concerned not only with the actual failure but with the intermittent stage after diagnosis of a failure and the change over to another unit. In other words, no loss of data is permitted, no operation errors are allowed, and no delay in the system reaction can be tolerated. From a system designer's point of view, these fault-tolerant requirements not only call for certain redundancy techniques but in many cases also for fully operative redundancy techniques. Simple redundancy techniques, although able to increase the system reliability, are not enough to achieve the required maintainability.

The cost of any technique for increasing system reliability is extremely high, because any redundancy drastically increases the system cost. The techniques suggested here are aimed at obtaining the maximum maintainability while keeping the system cost as low as possible.

14.2 REDUNDANCY TECHNIQUES

Redundancy is defined as the existence of more than one means for accomplishing a given task. The simplest redundancy technique is to connect an identical duplicate system in parallel with the existing system. This results in two entirely separate systems operating in parallel, where one of them acts as a 'spare'. The cost of such a double system is over double that of the cost of the single system, while the maintainability though improved is still far from perfect. Obviously this is not the ideal solution to be used in Command and Control systems.

Before discussing the redundancy techniques to be used, the various redundancy models must be evaluated. As mentioned before, consideration will be based on system modules, which, for this analysis only, may be considered to have an equal probability of failure. Each system module is assumed to operate as a separate unit and may consist of a number of functional circuits. For this discussion a number of terms must first be defined. R is defined as the complete system reliability, which is the probability of the successful operation of the system in performing its missions. $\bar{R}$ is the complement of the reliability factor, which is the probability of a failure occurring in the system. According to George Booles' basic law, the universe is defined as 1, giving in this case $R + \bar{R} = 1$, hence the reliability factor of a system may be realized as $R = 1 - \bar{R}$.

Accordingly, r and $\bar{r}$ are respectively designated as the success and failure of each individual module within the system. The module reliability must also confirm to the law of $r + \bar{r} = 1$.

Any system configuration generally could be divided into a series of n modules (as shown in Fig. 14.1). This configuration is known as a chained structure, since any failure

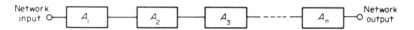

Fig. 14.1. Series system structure giving a mission reliability of $R_s = r^n$.

in a single module will result in a failure of the whole system. The mission reliability of the whole structure is equal to the multiplication of the individual reliability of each module:

$$R_s = r_1 \cdot r_2 \cdot r_3 \ldots r_{n-1} \cdot r_n.$$

For this discussion it is assumed that the reliability of all the individual modules is identical, and thus the following equation is received for a series module structure:

$$R_s = r^n.$$

Since the module reliability r is smaller than one, thus the more modules connected in series, the smaller R becomes.

The basic redundancy configuration is a parallel structure where each of the m modules performs the identical function (as shown in Fig. 14.2). The system will still

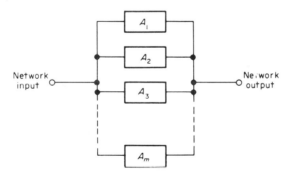

Fig. 14.2. Parallel system structure giving a mission reliability of $R_p = 1 - (1-r)^m$.

operate properly with multiple failures, provided that at least one unit is functioning. The probability that the whole system will fail is equal to:

$$\bar{R}_p = \bar{r}_1 \cdot \bar{r}_2 \cdot \bar{r}_3 \ldots \bar{r}_{m-1} \cdot \bar{r}_m.$$

As all the modules have the same reliability probability, all the units have the same probability of a failure, and thus the equation for a parallel module structure will be equal to:

$$\bar{R}_p = \bar{r}^m.$$

From the above equation one can calculate the mission reliability of a parallel equation by using the basic rule:

$$R_p \quad = \quad 1 - \bar{R}_p = 1 - \bar{r}^m$$

$$\text{since } \bar{r} \quad = \quad 1 - r,$$

$$\text{then } R_p = \quad 1 - (1 - r)^m.$$

The two equations received for a series and a parallel structure enable us to calculate the reliability mission in various redundancy configurations.

In a parallel-series combination (as shown in Fig. 14.3) each path consists of a series

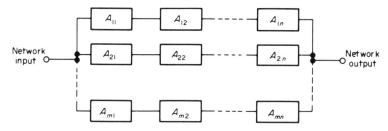

Fig. 14.3. Parallel–series system structure giving a mission reliability of $R_{ps} = 1 - (1 - r^n)^m$.

of n modules having m of these identical paths connected in parallel. This new structure is the same as the one already discussed for the parallel configuration, except that the system is not regarded any more as a single unit but is now divided into n modules. As before, it is assumed that all the modules in the structure have the identical reliability factor r. The mission reliability of each series path is equal to r^n, as already shown. The reliability of a parallel structure is also as shown before, i.e. $1 - (1 - r)^m$, except that the reliability r for each complete parallel path must be replaced by the reliability r^n representing the series of modules in each path. This gives the equation for a parallel structure:

$$R_{ps} = 1 - (1 - r^n)^m.$$

A completely different mission reliability can be obtained by using a series–parallel combination (as shown in Fig. 14.4). In this structure there are the same number of modules as in the parallel–series combination, although it presents a better mission reliability. Each parallel branch of this structure has the mission reliability of $1 - (1 - r)^m$. However, since all the parallel branches are connected in series, the system reliability is the result of the multiplication of each branch reliability n times, giving:

$$R_{sp} = [1 - (1 - r)^m]^n.$$

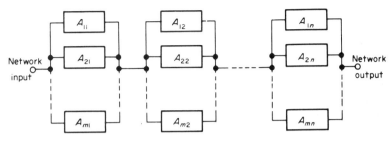

Fig. 14.4. Series–parallel system structure giving a mission reliability of $R_{sp} = [1-(1-r^m)]^n$.

There is no need for mathematical equations to realize that a series–parallel structure gives a much higher reliability factor than the equivalent system with a parallel–series structure. With series–parallel combination it is possible to create $m \times n$ different combinations, while with parallel–series it is possible to create only m combinations.

In practical applications it is not always possible to use an ideal series–parallel combination but only a mixture of the two types. Such a practical example is shown in Fig. 14.5, where both series–parallel and parallel–series are used. This combination is still much superior to a parallel–series structure.

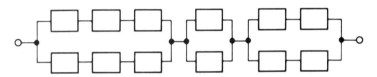

Fig. 14.5. Mixed parallel–series system structure giving a mission reliability of
$1-(1-r^3)^2] . [1-(!-r)^2] . [1-(1-r^2)^2]$.

To achieve any maintainability of system operation, some parallel redundancy must be used. However, the modules cannot be just directly connected in parallel, because a failure in one of the modules may "short circuit" all the other modules, consequently causing the failure of the whole structure. Two methods are in use for paralleling redundant modules, one using decision and the other switching elements.

Decision elements schemes are based on majority logic (as shown in Fig. 14.6), where at least half the modules must fail before the whole system fails. All the modules which are in parallel perform an identical function and all the outputs give the same value. The

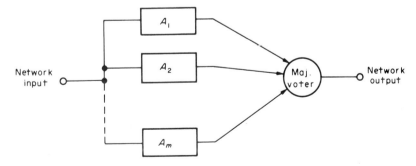

Fig. 14.6. A redundancy system based on majority voting logic.

majority logic checks all the outputs and presents its final output according to a majority vote; hence this logic is also termed a 'voter'. This majority logic has been extended to perform more complex functions, known as threshold logic, where the voter will present an output if k out of m modules present the same output. In other words, k out of m modules must fail in this redundant structure before the whole system fails. If the output does not yield the same answer from all the modules, the decision is taken by a threshold vote, where the threshold is defined by k. Obviously, the number of parallel modules m in a redundant structure must be greater than two and the threshold k must be greater than $m/2$. The most common majority logic is the triple-modular redundancy, referred to in the literature as TMR.

The advantage of majority voting in redundancy techniques is that no time is lost for changing modules, not to mention that there are no transient interruptions during failures. The mission reliability R of the voting construction is determined as before as a function of the reliability r of each module. The probability that the whole construction will not fail is either that none of the m modules fails or that not more than $(m-k)$ modules fail. This gives an approximate equation of the system reliability, which is equal to the sum of the two probabilities.

$$R = r^m + m\, r^{m-k}\, (1 - r)^k.$$

The above equation assumed that the majority voting circuit itself does not fail. For any useful application of the voting technique it must be warranted that the logic gate will be failure-free or at least more reliable than all the modules in parallel. In practice this requirement cannot always be guaranteed. Hence the voting circuit itself may be made redundant, as shown by the configuration given in Fig. 14.7.

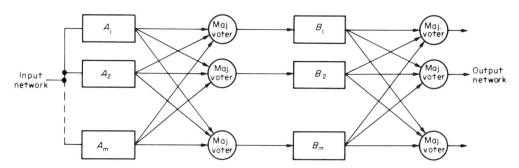

Fig. 14.7. A majority modular redundancy configuration.

Decision voting units in the form of k out of m are used for high reliability applications, such as for unmanned space ships. It should be pointed out that the reliability of each unit is reduced after each unrepaired failure. Thus, for extremely high reliability application, where the unit cannot be repaired, the redundancy requirement is that $(m-k)$ is larger than 1 and k is larger than m/k; for example, the redundancy configuration of 3 out of 5. In a 2-out-of-3 unmanned configuration, the failure of a single module will convert the system to a simple duplex system, with no apparent advantage to voting circuit while in a 3-out-of-5 configuration it still retains its voting power after the failure of one module. In a modular construction as shown in Fig. 14.7 a fault in a module will only affect one parallel branch and might be accepted in most applications. However, for unmanned spaceship application, either m and k must be increased or the complete triplicate voting unit must be triplicated. Another possible approach is to have spare modules which are switched in automatically when one of the basic modules fails, as shown in Fig. 14.8. This type of configuration has introduced further switching elements into the system, which raises the question of its reliability superiority.

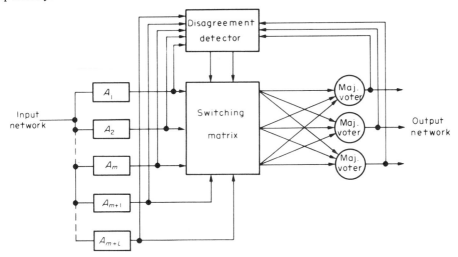

Fig. 14.8. A voter with m basic modules and l spare modules.

The main disadvantage of k out of m modules redundancy is that all the inputs and all the outputs are respectively connected, so that a failure in one module may cause the failure of the other modules. The first requirement then is that any failure in one of the modules will always result in presenting a negative output response, whereas only positive responses are required to activate the voter. It is also essential to require that these negative output responses, as a result of a failure, will not affect the other parallel modules.

Switching elements are based on standby redundancy, as shown in Fig. 14.9. In this

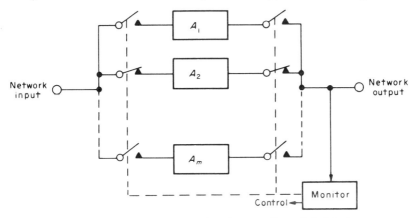

Fig. 14.9. A redundancy system based on standby switched logic.

type of redundancy only one of the parallel modules is connected to the system, while all the other modules are standing by. The monitor continuously checks the state of the output of the active module, and so, when a failure occurs, another module is switched in to replace it. Some systems may have switches only on the input to each module, some only on the output, while others may have switches to both the input and output. With all switching elements intermittent failures may be occurring while switching between the faulty module and the standby module.

In both decision and switching elements, the reliability of the monitor and the logic circuits (switching or voter) needs to be extremely high. As the number of redundant modules increases, the mission reliability of the parallel branch approaches the reliability factor of the monitor or that of the majority voter. When the reliability of the parallel modules is equal to that of the monitor or voter, then the parallel redundancy results in a mission reliability of a system which is considerably less than that of a single non-redundant module. Significant improvements in the system mission reliability can only be reached through the utilization of monitors and majority voters, which are much more reliable than the modules, or by duplicating and triplicating the monitor or majority voter logic circuits.

The redundant parallel modules may be in one of two forms, either active or standby. Active redundancy is used in a system where the redundant modules are continuously active and energized in performing the system tasks. In this category are modules performing identical tasks with the output being selected by majority voting. Another type of active redundancy that fits this definition is a system where the load tasks are shared between the parallel modules. Each module performs part of the system tasks, but when a failure occurs another module takes on an extra load. Standby redundancy is where the redundant modules do not perform any operational functions unless one of the primary modules fails. In this category, only the single operational module carries the full load, while all the standby modules are in reserve, waiting until a failure occurs. The standby modules may be fully or partially energized or completely inactive. The fully energized standby module may be only primed or may perform the same tasks as the active module.

Examples of both active and standby redundant systems have already been presented in the book. The parallel network in a multiprocessor configuration, as shown in Fig. 9.8, is an example of an active redundancy where the load is shared between the processors. A failure of one of the processors will not necessarily result in system malfunction, as the load of the faulty processor is transferred to the remaining operative processor. The duplicate computer system used in message-switching centres, as shown in Fig. 6.7, gives an example of a standby redundancy, where the standby module is fully energized and performs all the identical tasks of the active module.

Summarizing the discussion presented in this section, redundancy techniques can be classified by the following criteria:

(a) The system level at which the redundancy is applied, viz. whether a series–parallel or parallel–series redundancy is used.

(b) The state of the redundant modules, viz. whether the parallel modules are energized or inactive.

(c) The operational state of the redundant modules in performing the system tasks, viz. whether the parallel modules are active or standby.

(d) The portion of the tasks performed by each parallel module, viz. whether each module performs the full system load, part of the load or performs no tasks at all.

(e) The tasks performed by each parallel module, viz. whether identical tasks or different tasks (as in the case of shared loads).

(f) The type of monitor in the redundant construction in checking the output, viz. whether it uses switching elements or majority voting elements or a mixture of both as in a voter with spares.

14.3 PROTECTIVE REDUNDANCY

Reliability can be achieved by using reliable components where all the components are operated within their physical tolerances. The complex nature of command and Control system operation, however, cannot be guaranteed against failures by reliance on only the high reliability of its individual components and circuit. Higher reliability can be achieved only by means of redundancy which provides alternative facilities when failures are detected. Simple redundancy techniques, however, although providing alternative facilities, are no guarantee of maintainability. Maintainability can only truly be accomplished through the judicious use of protective redundancy techniques.

Protective redundancy is defined as the redundancy of hardware, software and of time. Hardware redundancy denotes additional circuits and functional modules. Software redundancy denotes additional programmes, while time redundancy denotes additional functions and operations in performing the specified tasks. A system contains protective redundancy if the faults can be tolerated by the system. The protective redundant measures are those which are not regularly required by the system in order to execute its specific tasks. Only when a failure occurs may some of the protective redundant measures be required to execute and carry on the system's specific tasks.

The various hardware redundancy techniques were discussed in the previous section, and this enables us to conclude a number of points for the design of Command and Control systems. The redundancy techniques introduced into the system design should not be placed on the system but on the module level. In other words, the system should be built in a modular form where the configuration is divided into the smallest possible operation units or modules, and where the redundancy is based on the duplicating or triplicating of these modules, as shown in Fig. 14.10. The modularity of a system indicates the degree by which each module can function independently, that is, for a given period of time without communicating with other modules. Another modularity requirement is that each module can be interconnected with other modules for the reconfiguration of the system upon the detection of a failure. The modules are so chosen that when a given module fails, it is easy to detect the error, diagnose the location and switch it out of the system.

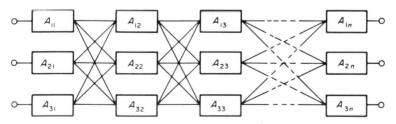

Fig. 14.10. System redundancy.

The reliability of an active parallel redundant configuration results in a higher reliability of the system than would be the case with that of a standby configuration. However, the effectiveness of the system, after a failure, may be less in an active than with a standby configuration. The reduction in system effectiveness may result in prolonging the required time duration for performing a specific task which may be encountered in a shared-load system. Another possibility of the system effectiveness being lowered is having to reduce the number of system facilities while keeping constant the operation time of the crucial facilities. For most Command and Control systems, active redundancy techniques with load sharing should be used, as this presents the best cost performance. Nevertheless, for systems where the performance effectiveness is critical, standby redundancy may be inevitable. Where ultra high reliability is required, resort must be had to majority voting.

With the design for maintainability, software redundancy must be applied to assist in the detection of a fault and the location of a faulty module. In this respect, redundancy means the use of redundant programmes which are not required for performing tasks. These programmes apply test diagnostic routines to detect faulty output on a logical operational combination. The repetitive frequency of applying these test programmes depends on the frequency of use of the individual unit or module. The diagnostic test programmes must be embedded into the system and operate automatically. These diagnostic measures, although performed by software operations, may be initiated by hardware measures.

Another software redundancy measure which should be adopted involves the use of error detection and/or error correction codes. Any data being transferred between the various units of the system, including the transfer of data to and from the memory and arithmetic unit, should be coded so as to allow the correction of any possible errors. Where the number of errors is higher than a specified criterion, it will indicate a failure located in the given module. The term redundancy, when used in this context in respect of error codes, denotes that not all the digits have an information meaning. The most common error-correcting codes used are those of Hamming, although in some systems cyclic codes are also used. In general, wherever data is transferred between units in the system, error checks should be introduced at both ends of the transmission path (including buses) to facilitate the diagnosis of faults. The advances in large scale integration (LSI) have reduced the cost of implementing error correction techniques and have accelerated the operations involved. Redundancy techniques, including error-correcting codes, increase the system reliability far beyond the state of the art of non-redundant systems.

Time redundancy is introduced by the interruption of the computer process in order to initiate a test programme. These programmes must have a high priority, to ensure that they are applied at the right frequency intervals. Further time redundancy is initiated after the location of a fault, and then a number of system operations must be repeated so as to reconstruct the failed process. Time redundancy must be allowed for locating the point of failure, and for reconstruction of the process in order to continue from there.

Protective redundancy must be so designed that the system may continue with its task in spite of hardware malfunctions. These protective techniques must provide total error control, and also the ability to detect the faults while the system is in operation. It must not only recognize and isolate the physical location of the fault but continue the operation from where it has failed. In some cases it may be required to go back a number of stages and to resume the process from there. It may also be required to cancel any further faults if the process had already continued before the original fault was isolated.

14.4 PRACTICAL CONSIDERATIONS

It should be obvious that high reliability can only be achieved by quadrupling, tripling or at least doubling the total number of modules in the system, but this requirement is regarded as unacceptable (in most commercial applications) for purely economical reasons. On the other hand, for all Command and Control systems it is essential that maintainability should be preserved despite the cost involved. In other words, for Command and Control systems it is imperative to apply some redundancy techniques, for which the acceptable minimum is a duplicated modular system. As hardware redundancy in itself is not enough to achieve the desired maintainability, the system must also apply both software and time protective redundancy techniques.

Duplication of the modules instead of triplicating or quadruplicating them imposes excessive design burdens on the fault tolerance of the system, since a double fault may cause a complete failure of the system. Above all other considerations the system design must ensure that the operation yields a good and fast diagnosis ability. The system must have a large number of test points which may easily be reached to permit continuous scanning, and each operation must have error detection media in order to detect the fault as soon as it occurs.

For the purpose of fault diagnosis, the duplicated modules of the same system must be regarded as a single construction. The two processors of the same system must be fully energized, while the standby processor carries out the identical tasks of the first processor even though its output may be blocked. Once an error is located in the active processor, the standby processor is automatically switched on-line. The failure in the first processor must be found with minimum delay, and this may be accomplished by having one processor diagnosing the other one. In other words, the standby processor may continuously diagnose the active processor, and after a failure the active processor (during any spare time) will diagnose the failed processor.

The above approach is not always satisfactory, and in many systems self-checking techniques are used, where each processor continuously diagnoses its own operation. The principle behind this is to reduce any cross-coupling between the active and standby processors and prevent any conflicting decisions between the results from the two units. Self-check techniques are dependent on hardware (assisted by software) aids which diagnose specific problems of the system operation. Checking techniques comprise test programmes, error codes and process simulation.

The system designer should appreciate that the implementing of the diagnosis techniques in an on-line environment adds more complexity to the maintenance of the system. It calls for the design of the system to be on three levels; fault tolerance, so that the operation will continue in all contingencies, fault location and fault repair. The implementation of the last item requires easy access to all locations, to enable each unit to be plugged in or out without affecting the systems on-line operation.

The maintainability design objective of Command and Control systems may be divided into two distinct groups, those which require to "fail-safe" in the presence of a malfunction and those which may be satisfied with only "fail-soft". A fail-soft system performs only the essentials of its tasks in the presence of a failure, while a system is said to be "fail-safe" if, in the presence of a failure, it can continue to perform its entire task load. Active redundancy schemes using majority voting, or standby redundancy schemes are basically fail-safe systems, while shared load redundancy schemes are intrinsically fail-soft.

Fail-soft offers a better cost performance, even though it reduces serviceability during a failure. Nevertheless, in systems where any degradation of service may cause extremely serious results (such as in medical hospital care) it is fail-safe techniques which must be used. The ideal fail-safe technique is achieved by active redundancy of k out of m. However, for practical and economical reasons most fail-safe Command and Control systems only use standby redundancy schemes. In other types of Command and Control systems, where reduction of services may be allowed, fail-soft schemes would be desirable.

Fail-soft schemes are characterized by the ability, in the presence of a failure, to reconfigure and allow the gradual degradation of services. These schemes may be designated partially-fault-tolerance, and as such are not suited for those Command and Control applications which require the continuation of the full task load. In the category of fail-soft schemes there are two classes of systems, those which share the work load between independent processors and those which parallel the essential modules. In the latter class only those parts which are needed for the real-time reaction must be duplicated. This technique is the cheapest, but it has the disadvantage that two programmes from the two modules may not be permitted to update a given file simultaneously. In practice a mingling of the two types is adopted. An example of a shared load is the multiprocessor.

In fail-soft schemes the work load is not constant, and the peak loads occur only rarely. The design objective of these systems must ensure that the system can cope with all the tasks when required to do so. As the essential service capacity of the system is far less than the peak, the redundancy need not cover all the modules but only those required for the essential services. The remaining modules, after the occurrence of a failure, must be able to diagnose the fault without interrupting the remaining system operation and so prevent a complete collapse. Self-diagnosis, using hardware and software techniques, must be used in each module separately.

Fail-safe, as distinct from fail-soft, can be designated a fault-tolerant system, as it has the ability to execute correctly all the system tasks regardless of any hardware failures. These fail-safe systems may be categorized into two classes, those using masking redundancy (such as majority voting) and those using standby redundancy. In masking redundancy the corrective action is wired-in and requires no further diagnosis. Masking redundancy, however, is much more expensive than standby redundancy, as it requires more modules in parallel. Furthermore, the standby redundancy principle attains better maintainability, fault detection and fault location. For this reason, standby redundancy seems to be better suited for fail-safe types of Command and Control applications. It must be noted that standby redundancy depends on the efficient design of the checking programme and the switch-over control unit.

Another method of categorizing fail-safe schemes is by static and dynamic redundancy. In static redundancy the spare modules are a permanent part of the system, while with dynamic redundancy there must be a switch-over process in the event of a failure. For most Command and Control systems applications, dynamic redundancy is the most effective technique both for cost and performance.

Switching, which is required for both fail-safe and fail-soft schemes, is the most unreliable feature in the system structure. The number of components in the change-over unit must be kept to a minimum and requires to be duplicated, since this is the most vital unit in the fault tolerant operation. The change-over unit must also provide synchronization between the various modules, to ensure smooth switch-over upon the detection of a fault. Intermittent changes during switch-over may be felt both in standby redundancy and in shared load, although less in masking redundancy. During these intermittent intervals data may be lost if no means are provided to prevent it. The effect of intermittent malfunctions during switch-over may be partially smoothed by suitable selection of the module size. The switch-over unit receives signals from the various functional modules which are activated under certain fault conditions. These signals cause an 'interrupt' on one of the other processors and can thus provide an automatic change-over.

14.5 COMMUNICATION MAINTAINABILITY ENHANCEMENT

The discussion so far has been concentrated on the maintainability of the central processor. This design approach is based on the fact that the central processor is the most expensive item in a system. Nevertheless, the maintainability of the system must consider all the aspects of the system and not only its processors.

The maintainability consideration of the communication network is that a connection must always be established and that when a failure does occur no data is lost. The probability of a communication link failure is far greater than one in a hardware processor; for this reason it is essential that there should be more redundancy inserted into the communication channels. The maintainability objective of communication, as achieved for the central processor, cannot be reached, since links will always be affected. Although the same effect of fault tolerance cannot be obtained, some means of fault detection and location must be introduced.

In the processor most of the faults are permanent, while in the communication network the faults are of an intermittent type. The system must be designed as fault tolerant in the presence of any intermittent non-permanent faults, such as those arising from the noise, delays and attenuation of the lines. In these intermittent faults, although all the equipment is operated properly and there is still a physical connection, the data transmitted is liable to be distorted. As before, the solution for these failures is of the type of protective redundancy using software and time redundancy. In communication, the software and time redundancy are interconnected, as most of the software actions require extra time. In fact, no transmission operation occurs without extra time being added for the software redundancy precautions. In this respect one can list the whole handshaking procedure in establishing the connection and in accepting the message. Further protective redundancy includes error control and the request for retransmission. These precautions are very effective for intermittent faults but are of no avail when the communication line is down.

The maintainability characteristics of a Command and Control system configuration depends on the topological layout of the network. The most common network is the centralized system. It has already been shown that the popular solution of star configuration is unreliable. Centralized systems, however, are unavoidable, and even in a distributed network each node acts as a local centre. Means must therefore be introduced to increase the probability of successful connection by moving away from the basic star structure.

A hardware fault in the communication network may be located in the terminal, modem, concentrator or in the communication line. The likelihood of a failure is mostly in the line rather than in the equipment. Furthermore, the equipment can be duplicated to increase its reliability but no great advantage can be gained by simply duplicating the communication lines, as shown in Fig. 14.11 (a). When adopting two or more links along the same given route, they inevitably pass along the same path, through the same cables and the same distribution exchanges. If there is a failure in one of the lines there is a great probability that there will be a failure also in the parallel redundant lines. A much better solution is an open-loop configuration, as shown in Fig. 14.11 (b), for a failure in one of the links still does not disconnect all the nodes beyond the failure point, since an alternative path could always be available. The loop shown in Fig. 14.11 (b) should not be associated with the loop network discussed in Chapter 8. Here the loop is open ended

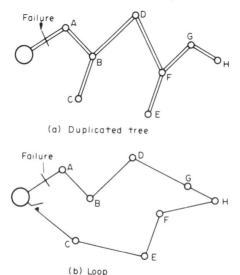

(a) Duplicated tree

(b) Loop

Fig. 14.11. Communication network redundancy. (a) Standby redundancy. (b) Protective redundancy.

and the direction of flow could be changed after a fault in the line has occurred. The interesting aspect of this solution is that the loop configuration is not only more reliable than the duplicated simple tree but is also much cheaper. When designing a centralized network the optimum reliability is obtained by means of several independent loops rather than a star network. It can be seen in Fig. 14.12 that the loop network shown in (b) provides less overall line mileage than the single star network shown in (a).

To summarize, one must conclude that the design objective of a communication network should be the finding of alternative routings rather than the duplicating of links. Where a network node is a concentrator, it should be provided with an automatic change-over which is operated when the first link fails. There is no need to provide the concentrators (or multiplexors) with sophisticated alternative routing tables, as there is no change in the node address and the switch-over could be operated as soon as the link is down. In distributed shared resources networks, a message can travel through a number of

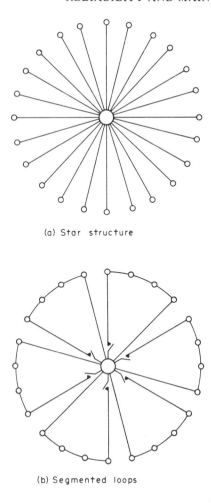

(a) Star structure

(b) Segmented loops

Fig. 14.12. Communication network configuration indicating that loops are cheaper and more reliable than star.

nodes before it reaches its final destination. Each node should be provided with a sophisticated table of alternative routes, but here too no duplicated lines should be used. In other words, it should be a matter of establishing the connection rather than planning for the shortest line between any two points. The maintainability ability of the network configuration to sustain communication depends on the number of completely independent and isolated paths between the nodes.

Fault detection and fault location must also be used in the communication network. The system must continuously establish the proper operation of all the links and all the nodes or terminals in the network. Periodic checks should be performed by the local centre to test the performance of all the lines and the equipment associated with this centre. The tests are created by a handshaking procedure where an affirmative answer must be received. These periodic tests should not only check the establishment of the proper connection but also measure the quality of transmission in the line, i.e. the signal-to-noise level. A major fault tolerance action to be taken in the presence of a failure is to inform all the other nodes in the network of any fault location so that they can adopt the necessary preventive actions, namely, to transfer to alternative routes when there is a link failure and, in the event of an equipment failure, to hold on to the message until the destination path becomes available again.

A point to be stressed is that the data must be retained despite any line failure. As soon as a destination is 'unavailable', the data should be stored for an indefinite period until the communication can be established again. No data may be erased before a positive acknowledgement is received from the other end. It is always advisable to inform the user that a message cannot reach its destination for some reason or other, because he might consider cancelling the message or require it to be routed to another destination. Each node which contains a front-end processor or a communication processor should be equipped with a secondary memory of sufficient capacity to cope with the overload of message traffic arising from the failure of some of its links. The front-end processors and the communication processors are also susceptible to failure and it is advisable to duplicate them locally.

14.6 REFERENCES

1. Von Alven, W.H. (Ed.), *Reliability Engineering*, Prentice Hall Inc., 1964, 593 pp.
2. Shooman, M.L., *Probabilistic Reliability: An Engineering Approach*, McGraw Hill, 1968, 524 pp.
3. Avizienis, A., "Design of fault tolerant computer", *AFIPS*, Vol. 31, Fall Joint Computer Conference, 1967, pp. 733-743.
4. Mathur, E.P. and Avizienis, A., "Reliability analysis and architecture of a hybrid redundant digital system: generalized triple modular redundancy with self repair", *AFIPS*, Vol. 36, Spring Joint Computer Conference, 1970, pp. 375-383.
5. Lyons, R.E. and Vanderkulk, W., "The use of triple-modular redundancy to improve computer reliability", *IBM Journal of Research and Development*, Vol. 6, No. 2 (Apr. 1962), pp. 200-209.
6. Chang, H.V. and Scanlon, J.M., "Design principles for processor maintainability in real-time systems", *AFIPS*, Vol. 35, Fall Joint Computer Conference, 1969, pp. 319-327.
7. Short, R.A., "The attainment of reliable digital systems through the use of redundancy – a survey", *Computer Group News*, Vol. 2, No. 2 (Mar. 1968), pp. 2-17.
8. Hsiao, M.V. and Tou, J.T., "Application of error correcting codes in computer reliability studies", *IEEE Trans. on Reliability*, Vol. R-18, No. 3 (Aug. 1969), pp. 108-118.
9. Dent, J.J., "Diagnosis engineering requirements", *AFIPS*, Vol. 32, Spring Joint Computer Conference, 1968, pp. 503-507.
10. Cole, F.B. and Bell, W.V., "Self repair techniques in digital systems", *AFIPS*, Vol. 32, Spring Joint Computer Conference, 1968, pp. 509-514.
11. Blaauw, G.A., "IBM system/360 multisystem organisation", *1965 IEEE International Convention Record*, Pt. 3, pp. 226-235.
12. Avizienis, A., "Fault-tolerant computing: an overview", *Computer*, Vol. 4, No. 1 (Jan./Feb. 1971), pp. 5-8.
13. Carter, W.C. and Bouricius, W.G., "A survey of fault tolerant computer architecture & its evaluation", *Computer*, Vol. 4, No. 1 (Jan./Feb. 1971), pp. 9-16.
14. Dieterich, E.J. and Kaye, L.C., "A compatible airborne multiprocessor", *AFIPS*, Vol. 35, Fall Joint Computer Conference, 1969, pp. 347-357.
15. Hansler, R., McAuliffe, G.K. and Wilkov, R.S., "Optimizing the reliability in centralized computer network", *IEEE Trans. on Communication*, Vol. COM-20, No. 3 (June 1972), pp. 640-644.

Name Index

Alexander, A. A.   93
Amoroso, F.   94
Amstutz, S. R.   157, 214, 232
Anderson, R. R.   172
Andres, R. K.   34
Andrews, H. R.   94
Arndt, F. R.   198
Aupperie, E. M.   214
Avi-Itzhak, B.   172
Avizienis, A.   300

Balcovic, M. D.   266
Ball, C. J.   131, 157, 214
Baudot, E.   61
Beardsley, C. W.   284
Becher, W. D.   214
Bell, W. V.   300
Ben Hiat, R.   198
Bennett, W. R.   93, 266
Birmingham, H. P.   34
Blàauw, G. A.   198, 300
Black, W. W.   198
Blane, R. P.   157, 214
Boan, T. H.   214
Bocchino, W. A.   34
Bouricius, W. G.   300
Brown, D. T.   266
Burton, H. O.   266
Byclen, J. E.   232

Carbato, F. J.   198
Carter, W. C.   300
Chaney, W. G.   99
Chang, H. V.   300
Chatelton, A.   53
Chien, R. T.   266
Chou, W.   157
Chu, W. W.   113
Coffman, E. G.   198
Coker, C. H.   172
Cole, F. B.   300
Colimeyer, A. J.   284
Comber, E. V.   284
Crowther, W. R.   214

Davey, J. R.   93, 266
Davis, C. J.   10
Davis, R. M.   214

Decker, H.   131, 214
Dell, F. R. E.   131
de Mareado, J.   198
Dennis, J. B.   198
Dent, J. J.   300
Desmonde, W. H.   198
Dieterich, E. J.   300
Doll, D.   113, 131
Doll, D. R.   131, 157
Donn, E. S.   284
Dudick, A. L.   198

Edwards, D. B. G.   232
Ellis, R. A.   198
Enslow, P. H.   284

Faber, D. J.   214
Fand, R. M.   284
Fano, R. M.   214
Feistel, H.   284
Fletcher, W.   53
Flynn, M.   198
Forney, D. G.   266
Foster, L. E.   53
Franco, A. G.   266
Frank, H.   157
Franklin, R. H.   93
Freed, A. M.   34

Gilbert, E. N.   157
Gould, G. T.   10
Gray, J. P.   157
Grobstein, D. L.   215
Grochow, J. M.   232
Gryb, R. M.   93

Hagleburger, D. W.   250-4, 266
Halier, U.   266
Hamming, R. W.   238, 247-50
Hamsher, D. H.   34, 53, 131
Hansler, R.   300
Hays, J. F.   172
Head, R. V.   198
Heart, F. E.   214
Henzel, R. A.   214
Hersch, P.   93, 157
Hilsman, W. J.   10

Hobbs, L. C.   231
Hsiao, M. V.   300

James, R. T.   93

Kahn, R. E.   131, 214
Kamman, A. B.   232
Karnaugh, M.   60, 93
Karp, D.   159
Kaye, L. C.   300
Kirstein, P. T.   215
Kleinrock, L.   198
Kohavi, Z.   266
Kropfl, W. J.   172, 173

Laliotis, T. A.   198
Larson, K.   214
Law, H. B.   93
Lender, A.   93
Lin, S.   266
Lucky, R. W.   93, 266
Lyons, R. E.   300

Machover, C.   232
Martin, J.   53, 93, 112, 131, 157, 172, 266
Martino, R. L.   34
Marzocco, F. N.   34
Mathison, S. L.   131
Matt, H. J.   266
Matteson, R. G.   113, 131, 214
McAuliffe, G. K.   300
McPherson, J. C.   214
Mihailovics, S.   232
Mills, D. L.   131
Moll, A. P.   232
Mouborgne, J. O.   279
Muench, P. E.   266
Murphy, D. E.   94
Muth, R. J.   94

Nast, D. W.   93
Newport, C. B.   113, 131, 214
Notley, J. P. W.   53
Nyquist, H.   62-3, 93

Olivier, G. M.   198
Ornestin, S. M.   214

Pack, C. D.   198
Patterson, N. W.   266
Peck, P. L.   284
Peters, B.   283
Petersen, H. E.   283
Peterson, W. W.   266
Pierce, J. R.   93, 172, 173
Podrin,   198
Pollard, J. R.   53
Progier, M.   266
Pyke, T. N.   157, 214

Reyes, R. V.   10
Rhodes, J.   93
Richards, R. K.   93, 131, 266
Riley, W. B.   198

Roberts, G. L.   214
Roberts, L. W.   93
Roiz, E. F.   157
Rose, D. J.   53
Rosenthal, C. W.   232
Rudin, H.   113, 131
Ryzlack, J.   113, 131, 214

Sackman, H.   198
Salz, J.   93, 266
Salzer, J. H.   198
Salzman, R. M.   231
Savage, J. E.   284
Scanlon, J. M.   300
Schneider, P.   53, 131
Schneider, R. H.   34
Seroussi, S.   157
Shannon, C.   39
Sheingold, D. H.   53
Sherman, D. N.   172
Shipton, J. E.   94
Shooman, M. L.   300
Short, R. A.   300
Skatrud, R. O.   283
Smith, N. G.   94
Spragins, J. D.   172
Stafferud, E.   215
Standeren, J.   232
Steiner   150-3
Steinkraus, L. N.   10
Sullivan, D. D.   266

Thompson, E. M.   53
Toffler, J. E.   94
Tou, J. T.   300
Townsend, M. J.   214
Turn, R.   283
Tymann, B.   157, 214

Uhlig, R. P.   215
Unold, R.   10

Vanderkulk, W.   300
Velasco, C. R.   266
Vernan   279
Von Alven, W. H.   300

Walden, D. C.   214
Walker, P. M.   131
Ware, W. H.   283
Watson, R.   198
Weiss, G.   10
Weissman, C.   283
Weldon, E. J.   93, 266
Wessler, B. D.   214
West, L. P.   172
Westcott, R. J.   93
Westhouse, R. A.   198
Wier, J. M.   94
Wilkov, R. S.   300
Williams, M. B.   93
Wolf, W. J.   93
Wright, E. P. G.   266

Abbreviations Index

ACK (Acknowledgement) 147, 236, 239
A/D (Analogue to digital converter) 38, 45, 46, 49, 219
AM (Amplitude modulation) 42, 53, 64, 66, 68, 70, 75
ARPA (Advance Research Project Agency) 209-12
ARQ (Automatic repeat request) 235-7
ATDM (Asynchronous time division multiplexing) 99, 106-10, 112-23, 144
ATS (Asynchronous time sharing) 144-5

BCC (Bit character count) 147
BST (British standard time) 202

C³ (Communication command and control) 1
CC (Communication controller) 205-7
CPU (Central processing unit) 33, 117, 205-6
CRC (Cyclic redundancy code) 257-66, 280-2
CRT (Cathode ray tube) 217-24

D/A (Digital to analogue converter) 47-48
DCS (Distributed computer system) 211, 214
DDD (Direct distant dialling) 89
DDT (Digital data transmission) 55-94
DPM (Differential phase modulation) 74
DPSK (Differential phase shift keying) 74
DSB (Double side band) 65
DSBSC (Double side band suppressed carrier) 65

ENQ (Enquiry character) 147-8
EOB (End of block) 76
EOM (End of message) 76, 129
EOT (End of transmission character) 147-8
ETX (End of text character) 147-8

Fc (Carrier frequency) 65, 67
Fm (Modulated frequency) 65, 67
FDM (Frequency division multiplexing) 41-42, 52-53, 96-97, 100-1, 102, 103, 112
FEC (Forward error control) 235, 237
FF (Flip Flop) 40, 49, 69, 264

FM (Frequency modulation) 65, 66-68, 76
FSK (Frequency shift keying) 66-68

IMP (Interface message processor) 209-11
ISSB (Independent single side band) 65-66

LED (Light emitting diodes) 229
LRC (Longitudinal redundancy checking) 245-6
LSD (Large screen display) 229
LSI (Large scale integration) 294

M (Modem) 55-57, 60-76
MAJ (Majority) 254, 289-91
MCL (Message character length) 241
MERIT (Michigan Education Research Triad) 211-13
MSU (Michigan State University) 211
MUX (Multiplexor) 41-44, 52-53, 95-114

NAK (Negative acknowledgement) 147-8, 236
NRZ (Non-return to zero) 58-61, 67, 68, 79
NRZI (Non-return to zero unipolar) 60

PAM (Pulse amplitude modulation) 37, 39, 44
PCM (Pulse code modulation) 34, 40, 44-46, 49, 60, 119
PDM (Pulse duration modulation) 38, 39
PM (Phase modulation) 53, 70-76
PPM (Pulse position modulation) 38, 40
PSK (Phase shift keying) 53, 70-76
PWM (Pulse width modulation) 38, 40

QAM (Quadrature amplitude modulation) 64-66, 74-76

R (Reliability mission) 286-9
RZ (Return to zero) 58, 67, 68

SDM (Space division multiplex) 49-52, 119
SOB (Start of block) 76
SOH (Start of heading character) 147-8
SOM (Start of message) 76, 128
SOT (Start of text) 129

SSB   (Single sideband)   65-66
SSBSC   (Single sideband suppressed carrier)
         65-66
STDM   (Synchronous time division multi-
         plex)   98-99, 107, 112
STX   (Start of text character)   147
SYN   (Synchronous character)   147

TDM   (Time division multiplex)   37, 38,
         42-45, 52-53, 98-101
TIP   (Terminal interphase processor)
         210-11

TMR   (Triple-modular redundancy)   290-1
TTY   (Tele-typewriters)   223
TV   (Television)   230-1

UM   (University of Michigan)   211

VRC   (Vertical redundancy checking)   246
VSB   (Vestigial sideband)   65-66, 74

WSU   (Wayne State University)   211

Subject Index

Acknowledgement (ACK)   147, 236, 239
Address   108, 122-3, 128-9, 162
Adjustable equalizers   88-89
Advance Research Project Agency Network
    (ARPA-NET)   209-12
Alternative routings   154-7, 169-70, 298-300
Amplitude modulation (AM)   42, 53, 64-66,
    68-70, 75
    double sideband (DSB)   65
    double sideband suppressed carrier (DSBSC)
    65
    independent single sideband (ISSB)   65-66
    quadrature amplitude modulation (QAM)
    64-66, 74-76
    single sideband (SSB)   65-66
    single sideband suppressed carrier (SSBSC)
    65-66
    vestigial sideband (VSB)   65-66, 74
Analogue to digital converter (A to D)   38, 45,
    46-49
Analogue transmission   35-53
Asynchronous time division multiplexing (ATDM)
    99, 106-10, 112-23, 144
Asynchronous time sharing (ATS)   144-5
Asynchronous transmission   76-78
Attenuation distortion   82-83, 87
Automatic repeat request (ARQ)   235, 237
Auxiliary memory   183-4
Availability   176, 186, 285

Backing memory   183
Batch processing   177-8
Baud   61
Bipolar pulses   58-60
Bits per second (bits/s)   61
Block codes   234, 238
Block length codes   239-41
Block parity   246
Block size   239, 257

Centre switching   116-31
Chain network   137
Circuit switching   118-21
Code efficiency   237
Code format   57-61
Codes
    block   234, 238
    convolution   234, 238, 250-7

cyclic   257-66, 280-2
    geometrical   246
    Hagelburger   250-4
    Hamming   250-4
    longitudinal   245-6
    parity   48, 244-7
    vertical   246
Coding quality   237
Command post   3, 6-7, 217-18
Communication controller   206, 211
Communication processor   207-9
Communication tree   150
Companders   82, 84-85
Concentrators   95-96, 101-6, 110-12, 223
Constraint length   238
Contention   144-9
Convolution codes   234, 238, 250-7
Cost effectiveness   8
Cross bar   49-52, 190-2, 229
Cross talk   50, 82, 84, 270
Cyclic redundancy codes (CRC)   257-66,
    280-2
Cyphering   278-82

Data collection   4
Data execution   4
Data transmission   59-94
Debit   60, 61, 64, 73, 74
Decision making   1, 11, 14
Delay distortion   82, 86
Demand   180
Differential phase modulation (DPM)   74
Differential phase shift keying (DPSK)   74
Digital data transmission (DDT)   55-94
Direct distance dialling (DDD)   89
Display systems   4, 217-32
Distance of code   238
Distortion (pulse)   40, 82
Distributed computer resources   197, 199-215
Distributed computer system (DCS)   211, 214
Distributed intelligence   221-4
Distributed switching network   123
Distribution on lines   81-88
Double side band (DSB)   65
Duobinary   60, 68-70
Duplex transmission   56-57, 139
Dynamic addressing   184
Dynamic memory allocation   185

Echo suppressor   82, 85
Entrance keys   273-5
Envelope delay distortion   82, 87
Equilizers   86-88, 139
Error control   233-66, 297
    convolution codes   234, 238, 250-7
    cyclic redundancy codes   257-66, 280-2
    geometrical codes   246
    Hagelburger codes   250-4
    Hamming codes   250-4
    parity codes   48, 244-7
    vertical redundancy checks   246
Eye pattern   63-64

Fail safe   295-6
Fail soft   187, 295-6
Fault detection   285
Fault diagnosis   285-6, 295
Fault location   285, 295
Fault repair   295
Fault tolerance   286
File security   275-7
Fixed reference phase detection   72
Flow charts   12, 25
Format pulse   57-61
Forward error control (FEC)   235
Frame   48
Frequency division multiplexing (FDM)   41-42,
    52-53, 96-97, 100-1, 112
Frequency modulation (FM)   65, 66-68, 70
Frequency shift keying (FSK)   66-68
Front end processor   117, 122, 194, 207
Full binary transmission   59

Gant chart   30
Gas discharge display   228-9
Geometrical codes   246

Hagelburger codes   250-4
Half-binary   59
Half-duplex   56, 139
Hamming codes   247-50
Hamming distance   238
Handshaking   89, 90, 200-1, 275
Hard copy   220-1
Hold and forward   106, 110-11
Human engineering   17, 218, 229
Human factors   14-20

Impulse noise   67, 82, 83-84
Independent single side band (ISSB)   66
Information rate   237
Input/output processor   206
Integration   27-30
Intelligent terminal   137, 219, 221-4
Interaction (between systems)   27-30
Interaction (man-machine)   14-20, 218
Interface message processor (IMP)   209-11
Interrupt   206
Inter-symbol interference   60, 62-64

Job scheduling   179-82

Karnaugh   60
Keying
    amplitude (AM)   42, 53, 64-66, 68-70, 75
    frequency (FSK)   66-68
    on-off   37, 64, 67
    phase (PSK)   53, 70-76

Large screen display (LSD)   221, 224-31
Light emitting diodes (LED)   229
Light pen   221
Line controller   51, 205-6
Lines
    dedicated   89-91
    dialled   89-90
    leased   89-90
    private   89
Liquid crystal display   229
Logs   124, 277-8
Longitudinal redundancy checking (LRC)
    245-6
Loop network   159-73, 192-3, 211, 212

Maintainability   285-300
Majority gate   254, 289-91
Manchester   59
Man-machine interaction   14-20, 218
Maximum length   256, 258-9
Memory   126, 178-9, 183-6, 188-93
Message address   108, 122-3, 128-9, 162
Message format   128-9, 162
Message heading   128-9, 162
Message switching   118-31
Michigan Education Research Triad (MERIT)
    211-13
Mission reliability   266-90
Modem   55-57, 60-76
Modular standby   125-6, 172, 291-2
Modular-2 (Mod-2) arithmetic   241-4
Modulation   52-53, 61-76
    amplitude (AM)   42, 53, 64-66, 68-70, 75
    differential phase (DPM)   74
    frequency (FM)   65, 66-68, 70
    phase (PM)   53, 70-76
    pulse   38
    pulse amplitude (PAM)   37, 39, 44
    pulse code (PCM)   38, 40, 44-46, 49, 60,
        119
    pulse duration (PDM)   38, 39
    pulse position (PPM)   38, 40
    pulse width (PWM)   38, 40
    quadrature amplitude (QAM)   64-66, 74-76
    quadrature phase (PM)   53, 70-76
Multi-address message   120
Multicomputer   186-7, 194-8
Multiconnection network   137-8
Multidrop   135-44
Multilevel formats   58, 59, 60, 63, 64, 71, 74-
    76
Multilevel transmission   71, 74-76
Multiplexing (MUX)   41-44, 52-53, 95-114
    asynchronous time division (ATDM)   99,
        106-10, 112-23, 144

frequency division (FDM)   41-42, 52-53, 96-97, 100-1, 112
space division (SDM)   49-52, 119
statistical   108, 109-10
synchronous time division (STDM)   98-99, 107, 112
time division (TDM)   37, 38, 42-45, 52-53, 98-101
Multipoint   135
Multiprocessing   178, 186-93

Natural wait   179
Network   115-18, 134-9
    architecture   134-9
    centre   115-31
    chain   137
    distributed switching   115
    loop   159-73, 192-3, 211, 212
    multiconnection   137-8
    multidrop   135-44
    multipoint   135
    non-switched   115
    star   115-16, 123, 134-7, 297-9
    switching   115-31
Node   133, 235
Non-return to zero (NRZ)   58, 67, 68, 79
Nyquist   61, 62-63, 88

On-line   9
On-off form   37, 64, 67

Packet switching   211
Paging   179, 183-5
Parity   48, 244-7
Performance scale   9
Pert diagram   30
Phase jitter noise   82, 84
Phase modulating (PM)   53, 70-76
Phase shift keying (PSK)   70
Plasma display   229
Point-to-point   89-90, 95, 135
Polling   144-9
Polynomial   243-4, 255-6
Priority   122, 179, 180-1
Privacy   267-84
Programmable concentrators   110-12
Protective redundancy   286, 293-5
Pseudoternary   60
Pulse distortion   40
Pulse formats   57-61
Pulse modulation   38
    pulse amplitude modulation (PAM)   38, 39
    pulse code modulation (PCM)   38, 40
    pulse position modulation (PPM)   38, 40
    pulse width modulation (PWM)   38, 40

Quadrature amplitude modulation (QAM)   64, 74-76
Quadrature phase modulation   59-60
Quadrature transmission   59-66, 74-75
Quantization   46
Quantum time   179, 180, 181, 182
Queueing   180-2

Random time sharing   149
Reaction time   9
Real-time   1, 9, 202
Recurrent code   147, 250-4
Redundancy   172, 233-4, 285-300
    active redundancy   292-3
    protective redundancy   286, 293-5
    shared load redundancy   292, 296
    stand-by redundancy   125-6, 172, 291-3
    triple-modular redundancy   290-1
Reliability   125-6, 140, 171-2, 196, 237, 285-300
Resource sharing   199, 215
Retransmission   235-7
Return path   235
Return-to-zero (RZ)   68
Ring counter   43-44, 257
Ring network   140, 143
Round robin discipline   181-2
Routing   129, 149-57
    alternative   154-7, 169-70, 298-300
    split   153-4, 159

Sample and hold   46-49
Scanning   44
Scheduling   179-82
Secrecy   201, 267-84
Security   201, 267-84
Selecting   145-9
Shanon's theorem   39
Shared computer resources   195, 201-3
Shared load redundancy   292, 296
Shift keying
    amplitude (AM)   42, 53, 64-66, 68-70, 75
    frequency (FSK)   60-68
    on-off   37, 64, 67
    phase (PSK)   53, 70-76
Simplex transmission   56
Simulation   26, 33
Single side band (SSB)   60, 65-66
Slots   160-1
Smart terminals   219, 221-4
Soft copy   220-1
Space division multiplex (SDM)   49-52, 119
Split routing   153-4, 159
Stand-by   125-6, 172, 291-3
Star network   115-16, 123, 134-7, 297-9
Start-stop transmission   76-78
Statistical multiplex   108, 109-10
Steiner's point   150-3
Steiner's tree   150-3
Store and forward   106, 110-11, 123
Super space multiplex   51-52
Supervisory control   179-82
Suppress side band   65-66
Swap time   180
Switching
    centre   116-31
    circuit   118-21
    distributed   115-16
    loop   159-73, 192-3, 211, 212
    packet   211

Synchronization   44, 76-81
Synchronous pulse   45, 48, 147-8, 162
Synchronous time division multiplex (STDM)
     98-99, 107, 112
Synchronous time sharing   144
Synchronous transmission   76, 78-81
Syndrome   253-4
System design   11-33
System stand by   125-72
System throughput   9

Telemetry   35-54
Terminal   10, 217-24
Terminal interphase processor (TIP)   210-11
Test programme   294-5
Threshold logic   290-1
Throughput   9, 238
Time accumulation   179
Time division multiplex (TDM)   37, 38, 42-45,
     52-53, 98-101
Time sharing   3, 9, 177-86

Time spectrum   42
Torn tape   110
Traffic controller   130
Triple-modular redundancy (TMR)   290-1
Transmission
     analogue   35-36
     data   59-94
     loop   159-73
Trunk signalling   82, 85
Turn-around time   143, 147-8

Unipolar pulses   58-59

Vernan technique   279
Vertical redundancy checking (VRC)   246
Vestigial side band (VSB)   66, 74-76
Virtual memory   184, 185
Voting logic   290-1

Wave form   46, 47, 48
White noise   82, 83